Joint Source-Channel Coding of Discrete-Time Signals with Continuous Amplitudes

Communications and Signal Processing

Editors: Prof. A. Manikas & Prof. A. G. Constantinides
(Imperial College London, UK)

Communications and Signal Processing – Vol. 1

Joint Source-Channel Coding of Discrete-Time Signals with Continuous Amplitudes

Norbert Goertz
The University of Edinburgh, UK

Imperial College Press

ICP

Published by

Imperial College Press
57 Shelton Street
Covent Garden
London WC2H 9HE

Distributed by

World Scientific Publishing Co. Pte. Ltd.
5 Toh Tuck Link, Singapore 596224
USA office: 27 Warren Street, Suite 401-402, Hackensack, NJ 07601
UK office: 57 Shelton Street, Covent Garden, London WC2H 9HE

British Library Cataloguing-in-Publication Data
A catalogue record for this book is available from the British Library.

ISBN-13 978-1-86094-845-9
ISBN-10 1-86094-845-6

Desk editor: Tjan Kwang Wei

Printed in Singapore.

To my wife Heike

Preface

This book is a result of both my research in the field of joint source-channel coding and my teaching in the field of Communications Engineering over the last decade at the Universities of Kiel, Munich, Lund and Edinburgh.

Joint source-channel coding is a field of ongoing research, and it can be seen as a part of the more general area of "Cross-Layer Design". Thus, this book is actually a snapshot that, of course, also reflects my personal preferences.

The book contains some more specialised chapters, which deal with "near-optimum joint decoding" (Chapter 3) and with good encoder designs for a number of special but practically relevant cases (Chapters 4–7). To make the material more accessible, I added the introductory Chapter 2 that provides the basic theory and an overview of practical approaches. In Appendix A I have collected results for the theoretical performance limits of communication systems. Although these limits of Information Theory are not the main topic of this book, they are still important as they form the ultimate limits for any practical system, which may or may not use joint source-channel coding. This collection of theoretical limits can be a useful reference and, hence, may also be of value to those not directly interested in the main topic of this book.

I would like to express my sincere appreciation in particular to Prof Joachim Hagenauer, Munich University of Technology, and to Prof Ulrich Heute, Christian-Albrechts-University Kiel, for giving me the opportunity to pursue the teaching and research for my "Habilitation"-project and for their sustained support and encouragement.

Special thanks go to Prof John B. Anderson, Lund University, for his advice, encouragement and lasting support.

Further I would like to thank all colleagues and friends who, although not personally mentioned, supported me in any way.

Last but not least, I would like to thank Katie Lydon from Imperial College Press for her support and Prof Athanassios Manikas, Imperial College London, for pointing me to the "Communication and Signal Processing Series" of World Scientific Publishing/Imperial College Press.

NORBERT GOERTZ

University of Edinburgh

Contents

Chapter 1

Introduction

The main goals of this book are to provide a theoretical framework for joint source-channel coding and to improve communication systems for the transmission of continuous-amplitude signals such as speech, music, images, or videos. Due to the desired practical implementation of the systems by digital computers the signals are represented in discrete-time, which is always possible by appropriate low-pass-filtering and subsequent sampling.[1]

Practical systems will always have constraints on the delay between the input of a signal sample and its output at the receiver. The tolerable delay is determined by the application and by complexity constraints. As an example, let us consider the transmission of speech in GSM mobile radio: the speech signals are band-limited to less than 4 kHz and sampled with a period of $T_A = 125$ μs (sampling frequency $f_A = 8$ kHz). The delay is critical in such an application since it strongly affects the quality of the conversation between two users. Therefore, it is limited to 20 ms, i.e., the maximum[2] number of samples that may be jointly processed is $N = 20\text{ms}/125\mu s = 160$. In non-conversational applications such as video streaming the delay constraints are much weaker, but still N is limited, possibly more because of memory limitations than by the requirements of the application itself.

For large N, information theory[3] implicitly suggests a way, of how to

---

[1]Since the word-length is also limited in digital computers, the raw signals after sampling are already quantized, but with high resolution; this allows to regard them as quasi-continuous.

[2]We have neglected in this discussion that in GSM the channel-coded bits of the source encoder are interleaved across two blocks, i.e., the truly occurring systematic delay is larger than 20 ms.

[3]In Appendix A, we summarize some results of information theory that are upper bounds for the best possible system performance, if there is no constraint on the block-length N. As these bounds are usually weak in our case of interest (we have to match

design a communication system. This approach is usually called the "separation principle." Roughly speaking it states that, for infinitely large block-lengths N, the transmitter can be separated into source and channel coding without loss in performance. The source encoder produces a sequence of independent and uniformly distributed bits (at a bit rate that equals the capacity of the channel) and the channel code protects the bits against the channel noise. Error correction is perfectly possible (in theory) if the code rate of the channel code (the number of data bits divided by the number of channel code bits) is not larger than the channel capacity. Except for the data bits, no additional communication between the source and channel encoders and decoders is required, as both components work perfectly.[4] Thus, the design of a communication system is greatly simplified, as, e.g., the design of a channel code is not influenced at all by the type of data at the input of the source encoder.

With the invention of turbo codes [Berrou and Glavieux (1996); Hagenauer *et al.* (1996)] and the developments based thereon, some promises of information theory have more or less become practice. For systems with large block-lengths, turbo codes are able to reduce the residual bit error rate after channel decoding down to 10^{-5} and less, at a channel quality that is close to the theoretical limits for error-free transmission.

If, however, the block lengths are moderate or even short (as, e.g., in mobile telephony), both the best known source and channel coding schemes don't even work "almost" perfectly. Hence, there are strong potential gains in transmission quality if the imperfectness of each system component is exploited by all others; that is why we deal with joint source-channel coding.

In Chapter 2 we give an overview over joint source-channel coding and summarize some approaches, especially for the decoder side. In Chapter 3 we state the theory of (near-)optimum joint source-channel *decoding* and an approximation called iterative source-channel decoding. We show that, in principle, it is possible find the optimal receiver for any given transmitter; the main issue in joint *decoding* is complexity reduction. In contrast to that, the general solution for the best transmitter for a given source and channel under given delay constraints is unknown. It is, however, crucial to deal with the encoder design, as even a perfect receiver for a badly chosen

---

delay constraints), we deliberately put this chapter into the Appendix. Analytical results for the best possible performances of delay-constrained systems are, unfortunately, widely unknown, so often the asymptotic information-theoretic results are the best we have for a comparison.

[4]One could also argue that this is the only "static" information the system components need to know about each other.

transmitter will lead to an overall bad system performance. Since a general solution is hard to obtain, we deal in the Chapters 4–7 with some special cases in which, due to additional practical constraints, "good" transmitter designs can be found.

Chapter 2

Joint Source-Channel Coding: An Overview

2.1 System Model

We will consider the basic model of a communication system that is depicted in Figure 2.1. The goal is to transmit the input source signal, the N-

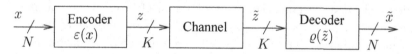

Fig. 2.1   System model

dimensional source vector x (e.g., N consecutive samples of a speech signal), by K channel-uses to a destination where we want to obtain a reproduction $\tilde{x}$ with the highest quality possible; the maximum block length N is limited due to delay constraints. The source vector components x_l, $l = 0, 1, ..., N-1$, have continuous amplitudes and we will use the mean squared error

$$d(x, \tilde{x}) = \frac{1}{N} \sum_{l=0}^{N-1} (x_l - \tilde{x}_l)^2 \tag{2.1}$$

as a quality measure for our reproduction. In what follows, we will assume that the vector dimension N and the number K of channel-uses are fixed and known and our goal is to find a deterministic encoder-decoder pair that minimizes the expected system distortion

$$D \doteq E_{X, \tilde{Z}}\{d(X, \tilde{X} = \varrho(\tilde{Z}))\} . \tag{2.2}$$

Where it is necessary, we will use capital letters to distinguish between random variables and their realizations; the latter we will be referred to by small letters. The probability density function (PDF) of the input vector

5

is assumed to be known and, for simplicity, we assume in this introductory discussion that adjacent source vectors are independent. For convenience, we will denote the PDF by $p(x)$ instead of $p_X(x)$ as long as this notation is unique. The random variables are re-introduced if the notation is ambiguous otherwise.

2.1.1 *Channel*

The transmission is carried out over a discrete-time channel, which is used K times to transmit the N source vector components; thus, the channel input signal in our model is the K-dimensional vector z. Where it is necessary, the input alphabet $\mathcal{Z}$ and the output alphabet $\tilde{\mathcal{Z}}$ of the channel are specified below. As an example, we may think of a K-bit channel input and a K-dimensional continuous-valued output from an additive white Gaussian noise channel. The model is, however, not restricted to channels with binary input, as illustrated by Figure 2.2. Moreover, the model can be extended such, that with one channel-use a vector of two real components is transmitted, i.e., the channel has two real dimensions in each of the K uses. This is important because frequently used modulation schemes like M-PSK[1] and QAM[2] are usually dealt with in the baseband representation, in which the signals are either regarded as two-dimensional real or, alternatively, as complex-valued.

We will assume that the noise on the memoryless channel is stationary and independent of all other signals and the components of the system. The conditional probability density function $p(\tilde{z} \mid z)$, which describes the probability density of the output vector $\tilde{z}$, given the input vector z, is assumed to be known. Since the channel is assumed to be memoryless, the PDF can be factored according to

$$p(\tilde{z} \mid z) = \prod_{l=0}^{K-1} p(\tilde{z}_l \mid z_l) \, . \qquad (2.3)$$

Usually, the input power of the channel has to be limited when the channel input alphabet is continuous. For instance, the power per channel-use, averaged over all vectors $z = \varepsilon(x)$, may be limited to P, i.e.,

$$\frac{1}{K} \, \mathrm{E}_X \{\|Z\|^2\} \le P \, . \qquad (2.4)$$

Such a power limitation is quite weak, because it does not limit the peak power of a channel-use. In other setups the average power for each channel

---

[1] M-ary phase shift keying
[2] quadrature amplitude modulation

input vector may be limited, i.e., $\frac{1}{K}\|z\|^2 \leq P$, or the power limit may apply to each of the K channel-uses,[3] i.e., $\|z_l\|^2 \leq P$. In the latter two cases the peak power *is* limited.

2.1.2 *Encoder*

The encoder, which is described by the unique deterministic mapping

$$z = \varepsilon(x) , \qquad (2.5)$$

with $x \in \mathcal{X} = I\!\!R^N$ and $z \in \mathcal{Z}$, is a device that maps the input source-signal vector x into the input z of the channel. The encoder output z may have to fulfill a constraint imposed by the channel (e.g., the power constraint mentioned above).

2.1.3 *Decoder*

The decoder is a deterministic device that maps the output $\tilde{z}$ of the channel to the decoder output signal, which should be a good estimate of what has been transmitted. The decoder mapping is denoted by

$$\tilde{x} = \varrho(\tilde{z}) , \qquad (2.6)$$

with $\tilde{x} \in \mathcal{X} = I\!\!R^N$ and $\tilde{z} \in \tilde{\mathcal{Z}}$. Since we want to obtain a good reproduction $\tilde{x}$ of the input x in the minimum-mean-square-error-sense, the output alphabet of the decoder equals that of the encoder input or it is a subset thereof. The alphabet $\tilde{\mathcal{Z}}$ of the channel output, however, can be different from $\mathcal{Z}$; as an example, one may consider the case of an AWGN channel with binary input.

The encoder/decoder design problem is illustrated by Figure 2.2 for the case of a real-input real-output channel and $N = 3$ source samples that are transmitted by $K = 2$ channel-uses.

2.2 System Distortion

We define the expected system distortion by

$$D(\varepsilon, \varrho) \doteq \mathrm{E}_{X,\tilde{Z}}\{d(X, \tilde{X}\} = \int\limits_{\tilde{Z}} \int\limits_{X} d\big(x, \tilde{x} = \varrho(\tilde{z})\big) \cdot p(\tilde{z}, x)\, dx d\tilde{z} , \qquad (2.7)$$

---

[3]This case will be considered in source-adaptive modulation in Chapter 6.

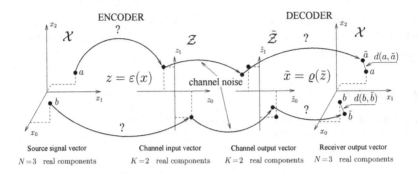

Fig. 2.2   Illustration of the encoder/decoder design problem for $N = 3$ and $K = 2$.

where $p(\tilde{z}, x)$ denotes the joint PDF of the random variables $\tilde{Z}$ (channel output) and X (encoder input). Clearly, both the encoder mapping ε and the decoder mapping ϱ are required to compute (2.7). In what follows, our goal is to choose both mappings such, that the expected distortion D is minimized. For this, it is useful to introduce an alternative representation of D: we may write

$$p(\tilde{z}, x) = p(x \mid \tilde{z}) \cdot p(\tilde{z}) . \tag{2.8}$$

By insertion of (2.8) into (2.7) we obtain

$$D(\varepsilon, \varrho) = \int_{\tilde{z}} D_D(\tilde{z}, \varepsilon, \varrho) \cdot p(\tilde{z}) \, d\tilde{z} , \tag{2.9}$$

with

$$D_D(\tilde{z}, \varepsilon, \varrho) \doteq \mathrm{E}_{X|\tilde{z}}\{d(X, \tilde{x})|\tilde{z}\} = \int_{\mathcal{X}} d\big(x, \varrho(\tilde{z})\big) \cdot p(x \mid \tilde{z}) \, dx . \tag{2.10}$$

The quantity defined in (2.10) is the conditional expected distortion, given a particular channel output $\tilde{z}$.

Similarly, we obtain

$$D(\varepsilon, \varrho) = \int_{\mathcal{X}} D_E(x, \varepsilon, \varrho) \cdot p(x) \, dx , \tag{2.11}$$

with

$$D_E(x, \varepsilon, \varrho) \doteq \mathrm{E}_{\tilde{Z}|x}\{d(x, \tilde{X})|x\} = \int_{\tilde{z}} d\big(x, \varrho(\tilde{z})\big) \cdot \underbrace{p(\tilde{z} \mid z = \varepsilon(x))}_{\text{cond. channel PDF}} \, d\tilde{z} , \tag{2.12}$$

where $p(\tilde{z} \mid x) = p(\tilde{z} \mid z = \varepsilon(x))$ was used, which is true, because the components in the system in Figure 2.1 form a Markov chain [McEliece (2002)], i.e., $\tilde{z}$ only depends on z and z only depends on x. The quantity defined in (2.12) is the conditional expected distortion, given a particular encoder input signal x.

2.3   Optimal Decoder for a Given Encoder

In a first step of the system optimization, our goal is to find the optimal decoder mapping $\varrho^{\circledast}(\varepsilon)$ for a given encoder mapping ε. Since the distance measure $d(x, \tilde{x})$ in (2.7) is non-negative for any combination of x and $\tilde{x}$, the distortion $D(\varepsilon, \varrho)$ is minimized, if $D_D(\tilde{z}, \varepsilon, \varrho)$ is minimized for each particular channel output vector $\tilde{z}$. Thus, the optimal decoder for a particular channel output $\tilde{z}$ is given by

$$\varrho^{\circledast}(\tilde{z}, \varepsilon) = \arg\min_{\varrho} D_D(\tilde{z}, \varepsilon, \varrho) \,. \tag{2.13}$$

With the mean squared error as a distance measure in the source signal space, the solution of (2.13) is the well-known mean-square estimator

$$\varrho^{\circledast}(\tilde{z}, \varepsilon) = \mathrm{E}_{X|\tilde{z}}\{X|\tilde{z}\} = \int_{\mathcal{X}} x\, p(x \mid \tilde{z})dx \,; \tag{2.14}$$

its derivation is given in Appendix B. By use of the Bayes-rule one can show that (2.14) equals

$$\varrho^{\circledast}(\tilde{z}, \varepsilon) = A \int_{\mathcal{X}} x \cdot \underbrace{p\big(\tilde{z} \mid z = \varepsilon(x)\big)}_{\text{cond. channel PDF}} \cdot \underbrace{p(x)}_{\text{source PDF}} dx \tag{2.15}$$

with

$$\frac{1}{A} \doteq p(\tilde{z}) = \int_{\mathcal{X}} p\big(\tilde{z} \mid z = \varepsilon(x)\big) \cdot p(x)\, dx \,. \tag{2.16}$$

For any deterministic choice of the encoder mapping ε, (2.15) and (2.16) define the optimal decoder which minimizes the expected system distortion for each particular channel output $\tilde{z}$. The optimality of the decoder mapping, however, does not guarantee a good system performance, because the encoder mapping might be a bad choice.

There is another interesting aspect: although the decoder is optimal for the given encoder, this does *not* mean that the performance of a system that uses this decoder cannot be better, if another encoder is used (clearly, there would exist an even better decoder if the encoder is changed). This topic is treated in some detail in Chapters 6 and 7.

2.4   Optimal Encoder

Now that we found the optimal decoder $\varrho^{\circledast}(\varepsilon)$ for any encoder ε (and any particular channel output $\tilde{z}$), we will use the result and try to optimize the encoder mapping to obtain an over all optimal system. Since our goal is to minimize the expected system distortion, the general optimization problem is given by

$$\varepsilon^{\circledast} = \arg \min_{\varepsilon:\, \text{constraint}} D(\varepsilon, \varrho^{\circledast}(\varepsilon)) \,, \qquad (2.17)$$

where the constraint in (2.17) is, for instance, the power limitation at the channel input mentioned above in Section 2.1. In most practically relevant situations the solution of (2.17) is intractable, even by numerical methods, but in some special cases solutions can be found as will be discussed later on. One problem is that, as we vary over the possible choices for ε, we have to vary the optimal decoder $\varrho^{\circledast}(\varepsilon)$, too. Hence, for simplification we may try to find an optimal encoder $\varepsilon^{\circledast}(\varrho)$ *for a given fixed decoder* ϱ; the corresponding optimization problem would read

$$\varepsilon^{\circledast}(\varrho) = \arg \min_{\varepsilon:\, \text{constraint}} D(\varepsilon, \varrho) \,. \qquad (2.18)$$

Although this optimization problem is still intractable in most cases, it is easier to solve than (2.17), because the decoder does not change within the optimization. The basic notion of this approach is to perform the system optimization iteratively: alternately, the decoder may be optimized for a fixed encoder and the encoder may be optimized for a fixed decoder.

The "constraint" in (2.17), (2.18) needs some elaboration. By use of (2.11), we may write the optimization problem as

$$\varepsilon^{\circledast}(\varrho) = \arg \min_{\varepsilon:\, \text{constraint}} \int_{\mathcal{X}} \underbrace{\int_{\tilde{\mathcal{Z}}} d\big(x, \tilde{x} = \varrho(\tilde{z})\big) \cdot p\big(\tilde{z}\,|\,z = \varepsilon(x)\big) \, d\tilde{z}}_{\overset{(2.12)}{=}\; \mathrm{E}_{\tilde{Z}|x}\{d(x, \tilde{X}) \,|\, x\} \;=\; D_E(x, \varepsilon, \varrho)} \cdot p(x)\, dx \,.$$

$$(2.19)$$

As above in the optimization of the decoder, the distance measure $d(x, \tilde{x})$ is non-negative for any combination of $x, \tilde{x}$. But if the *average* power of the channel input (per use) is constrained, i.e.,

$$\frac{1}{K} \mathrm{E}_X\{\|Z\|^2\} = \frac{1}{K} \mathrm{E}_X\{\|\varepsilon(X)\|^2\} \le P \,, \qquad (2.20)$$

we cannot optimize the encoder mapping for each particular input vector x, because the average power depends on the PDF of the source vector, which would not be considered in a minimization of $D_E()$ in (2.19).

If, however, the power $\|z\|^2 = \|\varepsilon(x)\|^2$ for the transmission of *each* particular source vector x is constrained, the optimization problem would read

$$\varepsilon^{\circledast}(x, \varrho) = \arg \min_{\varepsilon(x):\, \frac{1}{K}\|\varepsilon(x)\|^2 \le P} D_E(x, \varepsilon, \varrho) \,, \qquad (2.21)$$

i.e., we would optimize the channel input $z = \varepsilon^{\circledast}(x, \varrho)$ independently for each input source vector x. As we will see later in Chapters 6, 7 this optimization problem is tractable in practically relevant cases.

2.5 Special Cases

2.5.1 *Preliminary Remarks*

In this section some special cases are briefly discussed, for which optimal and realizable solutions of the joint source-channel coding problem are known. Further results on this topic have arisen in the recent literature [Gastpar *et al.* (2003)]. In contrast to these results, we will restrict ourselves to the practically relevant situation that a source and a channel are given, and a transmission system is to be designed that transmits the source signal over the channel with minimum mean squared error by use of deterministic mappings at both the encoder and the decoder.

2.5.2 *Gaussian Source and Gaussian Channel*

Let us assume that we want to transmit an independent discrete-time Gaussian source signal with the variance σ_x^2. The channel can be used exactly once per source sample, i.e., $N = K$, its average input power is limited to $\frac{1}{K}\mathrm{E}_X\{\|Z\|^2\} = P$, and the channel adds independent zero-mean Gaussian noise samples, which have the variance σ_w^2.

2.5.2.1 *System Example*

Figure 2.3 shows a possible transmission system [Berger and Gibson (1998)]. The individual samples from the source are transmitted independently, i.e., $N = K = 1$, and the encoder mapping has been chosen as

$$z = \varepsilon(x) = \frac{\sqrt{P}}{\sigma_x} \cdot x \,, \tag{2.22}$$

which is just a scaling of the input signal, in order to match the input power constraint of the channel.

The optimal decoder mapping can be found by evaluating (2.14):

$$\tilde{x} = \mathrm{E}_{X|\tilde{z}}\{X|\tilde{z}\} = \int_{-\infty}^{\infty} x\, p(x \mid \tilde{z})\, dx = \frac{1}{p(\tilde{z})} \int_{-\infty}^{\infty} x\, p(\tilde{z} \mid x) p(x)\, dx \,. \tag{2.23}$$

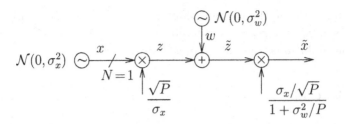

Fig. 2.3   Optimal system for the transmission of a uncorrelated Gaussian source over an additive-white-Gaussian-noise (AWGN) channel with average input power constraint.

Since the encoder mapping $z = \alpha \cdot x$, with $\alpha \doteq \sqrt{P}/\sigma_x$, is deterministic, we can equivalently write

$$\tilde{x} = \frac{1}{p(\tilde{z})} \int\limits_{-\infty}^{\infty} x \, p(\tilde{z} \mid z = \alpha \cdot x) p(x) dx \; . \tag{2.24}$$

The input signal and the channel noise are independent Gaussian random variables, i.e., the PDFs are[4] $p(x) = \mathcal{N}(0, \sigma_x^2)$ and $p(\tilde{z} \mid z) = p(w) = \mathcal{N}(0, \sigma_w^2)$. Since the channel output $\tilde{z}$ is the sum of the two independent Gaussian random variables Z and W, we have $p(\tilde{z}) = \mathcal{N}(0, P + \sigma_w^2)$. By insertion of the probability distributions into (2.24) we obtain for the decoder mapping (after same computations)

$$\tilde{x}(\tilde{z}) = \varrho^{\circledast}(\tilde{z}, \varepsilon) = \frac{\sigma_x/\sqrt{P}}{\left(1 + \sigma_w^2/P\right)} \cdot \tilde{z} \; . \tag{2.25}$$

The total system distortion is given by

$$D = \mathrm{E}_{X,W}\{\|X - \tilde{X}\|^2\} =$$

$$\int\limits_{-\infty}^{\infty} \int\limits_{-\infty}^{\infty} \left(x - \underbrace{\left(\frac{\sqrt{P}}{\sigma_x} \cdot x + w\right)}_{=\,\tilde{z}} \cdot \frac{\sigma_x/\sqrt{P}}{\left(1 + \sigma_w^2/P\right)}\right)^2 p(x)p(w)dxdw \; , \tag{2.26}$$

which equals

$$D = \frac{\sigma_x^2}{1 + P/\sigma_w^2} \; . \tag{2.27}$$

In terms of signal-to-noise ratio we obtain

$$\frac{SNR}{\mathrm{dB}} = 10 \log_{10} \frac{\sigma_x^2}{D} = 10 \log_{10}\left(1 + \frac{P}{\sigma_w^2}\right) \; . \tag{2.28}$$

[4]The notation $\mathcal{N}(\mu, \sigma^2)$ stands for the Gaussian PDF $p(x) = \frac{1}{\sqrt{2\pi}\sigma} \exp(-\frac{1}{2\sigma^2}(x - \mu)^2)$.

All results stay the same, if $N = K > 1$ is selected, because the independence of the input signal and the channel noise decouples the vector components in the evaluation of (2.14), so each vector component is transmitted separately. Hence, the decoder mapping (2.24) is optimal for any choice of N, even for $N \rightarrow \infty$. The optimality of the decoder mapping, however, does *not* guarantee that the encoder is optimal. Therefore, it is interesting to compare the performance of the system with the information theoretical bounds.

2.5.2.2 *Comparison with Results from Information Theory*

As summarized in Appendix A, the distortion-rate function of an independent Gaussian source signal with variance σ_x^2 reads

$$D(R) = 2^{-2R} \cdot \sigma_x^2 \qquad (2.29)$$

(R in bits/sample) and the channel capacity of the Gaussian channel equals

$$C = \frac{1}{2} \log_2 \left(1 + \frac{P}{\sigma_w^2} \right) \quad \frac{\text{bits}}{\text{channel-use}} \, . \qquad (2.30)$$

If we evaluate the distortion-rate function at the channel capacity, i.e., if we set $R = C$, we surprisingly also obtain (2.27), which means that the system in Figure 2.3 achieves the information theoretical bounds and, thus, no amount of source and channel coding of long blocks (large N) could improve the system [Berger and Gibson (1998)]. Therefore, the encoder mapping (2.22) is also optimal. This is a very special case, which is called "double matching of sources and channels" in [Berger (2003)]. The point is that the channel capacity is achieved with an Gaussian input signal, which is exactly the source signal in our setup—it is only scaled to match the power constraint.

This example shows that the inconsiderate interpretation of information theory and its uncritical use as a guideline for the design of a system can be misleading. The frequently used information theoretic approach for system design is the separation principle: a source encoder is applied that reduces the bit rate (that is required to represent the source signal) to the amount that can be transmitted error-free over the channel (channel capacity); the error-free bit rate is achieved by channel coding. If the signal has continuous amplitudes, source encoding is necessarily lossy, i.e., some distortion is introduced by source encoding (but not by channel noise, as it is the case in Figure 2.3). In order to work ideally, both source and channel coding need to process infinitely long blocks of source samples and bits.

Hence, we could have achieved the minimum distortion (2.27) by a system that uses the separation principle, but at the price of infinite delay and complexity—a bad choice from a practical point-of-view compared to the simple system in Figure 2.3.

The discussion above could evoke the illusion that we can forget about the separation principle and better should try to find simple "one-step" solutions as in Figure 2.3 also for other, practically more relevant combinations of sources and channels. This, however, turns out to be a bad idea again, because if *any* of the requirements for the source and the channel is not fulfilled (e.g., $N \neq K$, the source is correlated, or the channel noise is not Gaussian), the system in Figure 2.3 is no longer optimal and complex coding schemes with long block lengths *are* required to achieve a performance that is close to the information theoretical limits. This does *not* implicate that in practice we will apply the separation theorem in a strict sense. In fact, even for large block-lengths we will have to re-join source and channel coding in part to achieve good performance.

2.5.3   *Channels with Binary Input: Channel-Optimized Vector Quantization*

The problem (2.17) has been shown to be intractable in general, but for a channel with binary input (extendable to discrete inputs) and a relatively small number K of channel-uses per source vector, a numerical solution is known. The scheme described below is called channel-optimized vector quantization (COVQ) [Kumazawa *et al.* (1984); Zeger and Gersho (1988); Farvardin and Vaishampayan (1991)]; extensions can be found, e.g., in [Skoglund (1999); Alajaji and Phamdo (1998)].

As we assume a channel with binary input (e.g., a binary symmetric or additive white Gaussian noise (AWGN) channel with binary input), we have N source vector components that have to be mapped to K bits by the encoder. The set of possible channel-input vectors $z = \{z_0, z_1, ..., z_{K-1}\}$ is given by

$$z \in \mathcal{Z} = \{0,1\}^K . \tag{2.31}$$

Hence, the number of possible channel inputs is limited to

$$N_Y \doteq |\mathcal{Z}| = 2^K \tag{2.32}$$

and the K-bit input vector of the channel can also (more convenient) be represented by an integer number (index) $j \in \{0, 1, ..., N_Y - 1\}$, which we may relate to the bits z_n as follows:

$$j = \gamma^{-1}(z) \doteq \sum_{n=0}^{K-1} 2^n \cdot z_n \quad \text{with} \quad z_n \in \{0,1\}, \quad n = 0, 1, ..., K-1 . \quad (2.33)$$

This index is transformed back to the corresponding bit-vector by

$$z = \gamma(j) , \quad (2.34)$$

i.e., by inversion of (2.33).

Due to the limited number of possible channel inputs, the input source vector has also to be represented by a limited number N_Y of "values," i.e., x has to be *quantized*. This means that the source signal space $\mathcal{X}$ has to be partitioned into N_Y disjoint subsets $\mathcal{X}_j$, $j = 0, 1, ..., N_Y - 1$, with

$$\mathcal{X} = \bigcup_{j=0}^{N_Y - 1} \mathcal{X}_j . \quad (2.35)$$

Figure 2.4 illustrates the problem and the definitions above, for $N = 2$ continuous-valued source samples that are transmitted by $K = 3$ binary channel-uses.

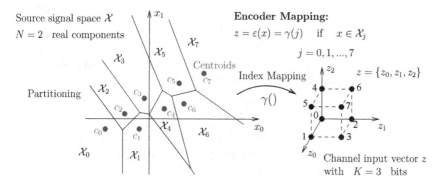

Fig. 2.4 Transmission of a two dimensional source vector x with continuous-valued components over a channel with binary input with $K = 3$ channel-uses.

If the partition regions $\mathcal{X}_j$ are known, which we will assume for the moment, the encoder is simply given by

$$z = \varepsilon(x) = \gamma(j) \quad \text{if} \quad x \in \mathcal{X}_j , \quad (2.36)$$

i.e., the K-bit channel input is the binary representation of the number j (see (2.33)) of the region $\mathcal{X}_j$, in which the source vector x is located.

As we want to design a transmission system that obtains a good reproduction at the decoder output in the minimum-mean-square-error sense, the optimal receiver mapping for the given encoder (2.36) can be found by application[5] of (2.15):

$$\tilde{x}^{\circledast}(\tilde{z}) = \varrho^{\circledast}(\tilde{z}, \varepsilon) = \mathrm{E}_{X|\tilde{z}}\{X|\tilde{z}\}$$

$$= \int_{\mathcal{X}} x \cdot p(x \mid \tilde{z}) dx = \frac{1}{p(\tilde{z})} \int_{\mathcal{X}} x \cdot \underbrace{p(\tilde{z} \mid x)}_{\text{channel PDF}} p(x) dx . \quad (2.37)$$

The integration over the whole region $\mathcal{X}$ can be split into a sum over the partition regions $\mathcal{X}_j$. As the encoder (2.36) is deterministic, $z = \gamma(j)$ can be equivalently used in the channel PDF instead of x, if $x \in \mathcal{X}_j$:

$$\tilde{x}^{\circledast}(\tilde{z}) = \frac{1}{p(\tilde{z})} \sum_{j=0}^{N_Y-1} \int_{\mathcal{X}_j} x \cdot p(\tilde{z} \mid z = \gamma(j)) \cdot p(x) \, dx . \quad (2.38)$$

This equals

$$\tilde{x}^{\circledast}(\tilde{z}) = \frac{\sum\limits_{j=0}^{N_Y-1} p(\tilde{z} \mid z = \gamma(j)) \int\limits_{\mathcal{X}_j} x \cdot p(x) \, dx}{\sum\limits_{j'=0}^{N_Y-1} p(\tilde{z}, z = \gamma(j'))} , \quad (2.39)$$

where the denominator has been expanded by a summation over all possible channel inputs. Finally, we obtain

$$\tilde{x}^{\circledast}(\tilde{z}) = \varrho^{\circledast}(\tilde{z}, \varepsilon) = \frac{\sum\limits_{j=0}^{N_Y-1} p(\tilde{z} \mid z = \gamma(j)) \cdot \mathrm{P}(j) \cdot c_j}{\sum\limits_{j'=0}^{N_Y-1} p(\tilde{z} \mid z = \gamma(j')) \cdot \mathrm{P}(j')} \quad (2.40)$$

with

$$\mathrm{P}(j) \doteq \int_{\mathcal{X}_j} p(x) \, dx \quad (2.41)$$

and

$$c_j \doteq \frac{1}{\mathrm{P}(j)} \int_{\mathcal{X}_j} x \cdot p(x) \, dx . \quad (2.42)$$

---

[5]We assume here that the source samples or at least the consecutive source vectors are uncorrelated.

The quantities c_j given by (2.42) are known as the "centroids" of the partition regions $\mathcal{X}_j$; these regions have not been discussed so far: for the moment, they are simply assumed to be given, which also includes their numbering (i.e., the choice of the bit-vectors for the binary transmission of the region numbers). The probability that a source signal vector lies within the partition region $\mathcal{X}_j$ is denoted by $P(j)$. It should be noticed that c_j and $P(j)$ can be computed and stored in advance, because both quantities depend only on the PDF of the source vectors and the partitioning of the source signal space; both are fixed when the transmission system is used. Hence, the partitioning (2.35) of the source space and the definition of the quality measure completely determine the encoder and the decoder (2.40); the latter is optimal for any choice of the partition regions[6] (encoder mapping).

The implementation of the encoder (2.36) is a problem in practice: in principle one could store mathematical descriptions of the regions $\mathcal{X}_j$, but this becomes infeasible if the source vectors have high dimensions. The problem can be solved as follows: we have assumed a channel with binary input; if we additionally assume antipodal signaling on the channel, a power constraint at the channel input is simply fulfilled by an appropriate choice of the signal amplitude for the bits. Thus, we can minimize the expected system distortion (2.11) by minimization of the expected "encoder distortion" (2.12)

$$D_E(x, \varepsilon, \varrho^{\circledast}(\varepsilon)) \doteq \int\limits_{\tilde{z}} \underbrace{d\big(x, \varrho^{\circledast}(\tilde{z}, \varepsilon)\big)}_{\text{mean squared error}} \cdot \underbrace{p\big(\tilde{z} \mid z = \varepsilon(x)\big)}_{\text{cond. channel PDF}} \, d\tilde{z} \qquad (2.43)$$

for a particular source input vector x, where we assume that in general the channel outputs can take any real number.

In order to find an alternative way for encoding, which avoids the direct use of the encoder mapping (2.36), we exploit that there is only a limited number of possible outcomes at the encoder output, i.e., $\gamma^{-1}(\varepsilon(x)) = j \in \{0, 1, ..., N_Y - 1\}$ for any x. Thus, we use a modified version of the expected encoder distortion, assuming that the i-th partition region has been selected by the encoder:

$$D_E(x, i, \varrho^{\circledast}(\varepsilon)) = \int\limits_{\tilde{z}} d\big(x, \tilde{x}^{\circledast} = \varrho^{\circledast}(\tilde{z}, \varepsilon)\big) \cdot p\big(\tilde{z} \mid z = \gamma(i)\big) d\tilde{z} \,, \qquad (2.44)$$

---

[6]To avoid any discussion about the validity of this statement, we will demand that the partition regions are convex, which is the only reasonable choice if the mean-squared-error distance measure is used.

with $i = 0, 1, ..., N_Y - 1$. In (2.44) the partition regions $\mathcal{X}_j$ are still required to compute the optimal decoder output $\tilde{x}^\circledast = \varrho^\circledast(\tilde{z}, \varepsilon)$ by (2.40) (due to (2.41) and (2.42)), but for the moment we will assume, that $P(j)$ and c_j, $j = 0, 1, ..., N_Y - 1$, have been determined "somehow," possibly by some codebook training procedure (this will be discussed later on). Now, to encode the input source vector x we perform the minimization

$$\varepsilon(x) = \arg\min_i D_E(x, i, \varrho^\circledast(\varepsilon)) \,, \tag{2.45}$$

which simply means that we compute $D_E(x, i, \varrho^\circledast(\varepsilon))$ for each possible index i and pick the one for transmission that gives the lowest distortion. By use of this concept, we may also write the partition regions as

$$\mathcal{X}_j = \{x \in \mathcal{X} : D_E(x, j, \varrho^\circledast(\varepsilon)) \le D_E(x, i, \varrho^\circledast(\varepsilon)), \ i = 0, 1, ..., N_Y - 1\} \,. \tag{2.46}$$

2.5.3.1   *Binary Symmetric Channel (BSC)*

If the channel output is also binary, only a limited number of K-bit channel *outcomes* $\tilde{z}$ exists. Thus, the optimal decoder will also generate only $N_Y = 2^K$ outputs; the latter are called COVQ code-vectors. They can be pre-computed by (2.40) and stored in a table—the so-called COVQ codebook. We will denote the code-vectors by

$$y_i \doteq \varrho^\circledast(\tilde{z} = \gamma(i), \varepsilon) \,, \quad i = 0, 1, ..., N_Y - 1 \,. \tag{2.47}$$

Thus, if some $\tilde{z} = \gamma(i)$ appears at the channel output it can be decoded by a simple table-lookup which will give y_i.

For the modified encoder according to (2.45), the integration in (2.44) is turned into a summation over the possible channel outcomes according to

$$D_E(x, i, \varrho^\circledast(\varepsilon)) = \sum_{j=0}^{N_Y-1} d\big(x, \underbrace{\varrho^\circledast(\gamma(j), \varepsilon)}_{\doteq\, y_j}\big) \cdot P(\tilde{z} = \gamma(j) \mid z = \gamma(i)) \,, \tag{2.48}$$

with $i = 0, 1, ..., N_Y - 1$. For a binary symmetric channel with bit error probability p_e, the "index" transition probabilities in (2.48) are given by

$$P_{j|i} \doteq P(\tilde{z} = \gamma(j) \mid z = \gamma(i)) = p_e^{d_H(i,j)} \cdot (1 - p_e)^{K - d_H(i,j)} \tag{2.49}$$

where K is the number of bits required to encode the indices by (2.34) and $d_H(i, j)$ is the Hamming distance between the bit-vector representations of transmitted index i and the received index j, i.e., $d_H(i, j)$ equals the number of bit errors inserted by the channel.

If we insert (2.49) into (2.48) we obtain the so-called COVQ distance measure

$$d_{\text{covq}}(x, y_i) = \sum_{j=0}^{N_Y-1} d(x, y_j) \cdot P_{j|i} , \quad i = 0, 1, ..., N_Y - 1 , \qquad (2.50)$$

where we have introduced the abbreviation $d_{\text{covq}}(x, y_i) = D_E(x, i, \varrho^{\circledast}(\varepsilon))$, and $d(x, y_j)$ is the mean squared error between the input vector x and the COVQ code-vector y_j. We will use (2.50) later on in Chapter 4.

2.5.3.2 Channels with Binary Input but with Real Output

If the channel has real outputs, we face another practical problem in the computation of $D_E(x, i, \varrho^{\circledast}(\varepsilon))$ by (2.44): in most cases, there is no analytical solution for the integral. Thus, we have to approximate the integration by an appropriate summation (similar as in [Skoglund (1999)]): for each possible K-bit vector channel input $z = \gamma(i)$ we perform a measurement of a large number, say N_i, of K-dimensional vector channel outputs $\tilde{z}$; the measured individual channel output vectors $\tilde{z}_k^{(i)}$ are grouped into the sets

$$S^{(i)} \doteq \{\tilde{z}_0^{(i)}, \tilde{z}_1^{(i)}, ..., \tilde{z}_{N_i-1}^{(i)}\} . \qquad (2.51)$$

We can compute an approximation of the integral in (2.44) by averaging the distortions over all measurements in the sets:

$$D_E(x, i, \varrho^{\circledast}(\varepsilon)) \approx \frac{1}{|S^{(i)}|} \sum_{\tilde{z} \in S^{(i)}} d\left(x, \tilde{x}^{\circledast} = \varrho^{\circledast}(\tilde{z}, \varepsilon)\right) . \qquad (2.52)$$

Thus, (2.52) can be used in (2.45) instead of (2.44), and $\varrho^{\circledast}(\tilde{z}, \varepsilon)$ for any channel output $\tilde{z}$ follows from (2.40). A major disadvantage of this approach is that we have to store the sets (2.51) at the encoder. A possible solution [Alajaji and Phamdo (1998)] for this problem is to (scalar) quantize the soft-outputs $\tilde{z}$ of the channel by M levels. Then, similar as for the BSC, we obtain a limited number $N_Y' = 2^{K \cdot M}$ (which can be much larger than the number 2^K of possible channel input bit-vectors) of estimates from (2.40), which can again be pre-computed and stored as the COVQ codebook.

For the computation of $\varrho^{\circledast}(\tilde{z}, \varepsilon)$ by (2.40), P(j) and c_j are required in any case. In what follows, we will show how to find both quantities by a numerical approach.

2.5.3.3   *Optimization of the Encoder Mapping: Codebook Training*

The encoder mapping (2.36) (and the associated optimal decoder mapping (2.40)) are completely specified by the partitioning of the source signal space according to (2.35) and the choice of the quality measure for the reproduction, which is usually the mean squared error. To achieve the best possible performance, the partitioning of the source signal space has to be optimized: this is the goal of "codebook training."

As already discussed above, the mathematical representation of the partition regions is hard to find (if it is possible at all) and hard to handle; that was the reason, why the encoding is practically performed by (2.45) and (2.50)/(2.52) instead of (2.36). In (2.40), $\varrho^{\circledast}(\tilde{z}, \varepsilon)$ is computed for which we need $\mathrm{P}(j)$ and c_j. Unfortunately, the partition regions are required to compute $\mathrm{P}(j)$ by (2.41) and c_j by (2.42). Moreover, the PDF of the source signal vectors is also required, but in most practical cases it is not known explicitly. Hence, there is no way to find an analytical solution for the optimization of the encoder mapping. The way out of this dilemma is a numerical approach whose concept is similar to the well-known LBG algorithm [Linde *et al.* (1980)] for source-optimized VQ codebook training. The procedure is described in [Farvardin and Vaishampayan (1991)] for a BSC, but with the definition of the expected distortion by (2.52) it also works for channels with binary inputs and continuous outputs. The training algorithm is described below:

COVQ Codebook Training:

(1) Initialization:

  - Take a training set $T = \{x_0, x_1, ..., x_{N_T-1}\}$ of source vectors x_l.
  - Specify the initial centroids c_j, $j = 0, 1, ..., N_Y - 1$, and their probability distribution $\mathrm{P}(j)$. (A source-optimized codebook trained by the LBG algorithm [Linde *et al.* (1980)] is a reasonable choice.) Set the "old" distortion $D_{\mathrm{old}} = \infty$.

(2) Encode the training data set by (2.45) and (2.50)/(2.52) and group the training vectors that are encoded to index j into the set T_j. Compute the expected total system distortion D_{new} for the current iteration by averaging over all encoding distortions:

$$D_{\mathrm{new}} = \frac{1}{N_T} \sum_{\tau=0}^{N_T-1} \min_i D_E(x_\tau, i, \varrho(\varepsilon)) \, . \qquad (2.53)$$

(3) Compare the total system distortion of the current iteration with the value from the previous iteration. If the relative change falls below a specified level (e.g., $(D_{old}-D_{new})/D_{new} < 10^{-3}$), then stop. Otherwise, set $D_{old} = D_{new}$; go to Step (4).

(4) Compute approximations for (2.41) and (2.42) according to

$$P(j) \approx \frac{|T_j|}{|T|} \quad \text{and} \quad c_j \approx \frac{1}{|T_j|} \sum_{x \in T_j} x \tag{2.54}$$

and compute new COVQ code-vectors by (2.40). Go to Step (2).

After convergence of the training procedure, a locally optimal partitioning of the source signal space and an index mapping (see Figure 2.4) are found. The choice of the initial centroids is, however, a critical issue: a reasonable way to do the initialization is to start with a source-optimized codebook derived from the LBG-algorithm [Linde *et al.* (1980)]. Then, the bit error probability (or the channel noise variance in the continuous case) is increased stepwise, i.e., the whole COVQ training procedure is repeated several times, in which the results of the preceding steps are used as initializations. As the convergence to a bad local minimum is a significant problem, it has been reported in [Heinen (2001)] that other schemes (source-optimized channel codes) can achieve better performance, although COVQ is conceptually optimal for the given limitations of the block lengths (N, K) as it imposes no further restrictions.

It has been observed in [Farvardin and Vaishampayan (1991)], that the training procedure described above can generate empty coding regions, i.e., some indices are never selected by the encoder. This is implicit error control coding. What happens is that the partition regions (see Figure 2.4) are deformed by the training process. Due to the channel error probabilities, the border lines between the partition regions which, in the pure source coding case, are perpendicular to lines connecting the centroids, are no longer perpendicular, and it may happen that some regions do not contain any training vectors at all after some iterations of the training procedure. As this is known at the encoder, these indices do not need to be checked in the distance computations (e.g., (2.48)) so computational complexity is saved at the encoder; the latter is especially true if the channel noise is strong. In [Farvardin and Vaishampayan (1991)] it is reported, that the number of used coding regions is smaller than half of the total number of possible channel inputs at a bit error rate of 0.1. It should be noticed that due to the channel noise still *all* possible vectors occur at the channel output, regardless of the fact that some bit combinations are never transmitted. Even in this case the optimal decoder (2.40) computes the best mean-quare estimate of the encoder input signal.

It has been shown in [Farvardin and Vaishampayan (1991)] and [Goertz and Kliewer (2003)] that COVQ can be implemented with no additional complexity or memory requirements (compared to conventional VQ), if the channel noise is stationary. But if the channel is time-varying, COVQ requires to store several codebooks (which cover the range of possible noise variances on the channel), which makes the scheme unattractive in those cases. In Chapter 4 we introduce a scheme which approximates optimal COVQ on time-varying channels, while the memory requirements are almost the same as in conventional VQ.

Due to the complexity of encoding and the memory requirements for the codebook, COVQ but also conventional VQ are only applicable if the number K of bits per source vector is small, e.g., $K \leq 10$. Therefore, COVQ cannot be directly used, e.g,. for speech coding, because at a bit rate of 1 bit per sample (typically required in mobile radio) the maximum block-length would only be $N = 10$ samples: unfortunately, source coding at that rate with sufficient quality is only possible with block lengths of a few hundred samples. Nevertheless, COVQ is a very interesting concept, because it is the only non-trivial scheme, in which the encoder mapping is *not* separated into partial mappings.

2.5.3.4   *Some Simulation Results for COVQ*

Some results for COVQ are given for an uncorrelated Gaussian source, which is transmitted over a binary symmetric channel with a bit error probability p_e at a rate of 1 bit per source sample (i.e., $N = K$). Figure 2.5 shows the results. The COVQ codebooks are always optimally matched to the channel, i.e., a new codebook was trained for each value of p_e. The optimal performance theoretically attainable (OPTA, for details, see Appendix A) is also included in Figure 2.5. The curve follows from the SNR-version of the distortion-rate function of a Gaussian source

$$SNR/\text{dB} = 6.02 \cdot R \,, \tag{2.55}$$

which is evaluated at $R = \frac{K}{N} \cdot C$, where

$$C = 1 - H(p_e) \tag{2.56}$$

is the channel capacity, with the binary entropy function

$$H(p_e) = -p_e \log_2(p_e) - (1 - p_e) \log_2(1 - p_e) \quad . \tag{2.57}$$

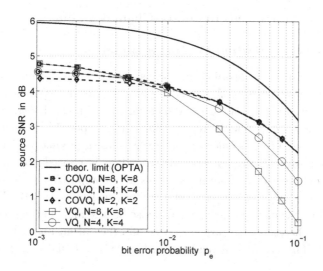

Fig. 2.5 Simulation results for COVQ over a BSC.

As in this case $K = N$, the result equals

$$SNR/\mathrm{dB} = 6.02 \cdot (1 - H(p_e)) . \qquad (2.58)$$

Figure 2.5 shows that the performance of COVQ gets better as the vector dimension increases. For $N = 8 = K$, however, there is still a large gap to the theoretically optimal performance (OPTA). Although eventually the OPTA is achieved by COVQ for $N \to \infty$, it can not be reached in practice due to complexity. Moreover, if due to a delay constraint the vector dimension is strictly limited, COVQ *is* the optimal scheme (supposed the training procedure for the codebook has converged against the global optimum) and the gap to the OPTA simply shows the unavoidable loss in performance due to the delay constraint.

The performances for purely source-optimized VQ have also been included in Figure 2.5: the plots show that COVQ is always superior to VQ, but for small bit error rates the curves coincide. The gain of COVQ over VQ is larger for large vector dimensions.

2.6   Practical Approaches to Source-Channel Coding

2.6.1   *Systems for Multimedia Transmission*

Consider the problem of encoding and transmitting a multimedia source signal vector[7] that consists of N_u source samples, e.g., from a speech signal. Due to delay constraints, the vector dimension N_u is limited; depending on the application (e.g., telephony) the limitation is possibly strong. In advanced multimedia codecs (such as code-excited linear prediction (CELP) [Schroeder and Atal (1985)] for speech), encoding is carried out in two steps: first, each vector u of source samples is decomposed into a set of parameter vectors[8] $x^1, x^2 \ldots$. A parameter vector consists, e.g., of the LPC-coefficients,[9] but the mean power of u could also be a (scalar) parameter. In the second step the vectors x^m are quantized and the output bit-vectors b^m are generated. This scenario is depicted in Figure 2.6. Usually, a channel

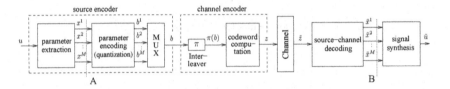

Fig. 2.6   Model of a multimedia transmission system.

code is used to improve the performance of the system in presence of noise on the channel. The channel codewords z are computed from the bit-vector b that is generated from the output bit-vectors $b^1, b^2, \ldots$ of the source encoder by the multiplexer.

In what follows, we will still consider the basic model of a communication system that is depicted in Figure 2.1, but now we will interpret the parameter vectors in Figure 2.6 as the source signals, i.e., we will plug-in the system in Figure 2.1 between the points "A" and "B" in Figure 2.6. For brevity of notation, we will only consider one parameter signal below (therefore we will omit the superscripts). This means that the mean-squared error of the source-codec *parameters* is used a quality measure and

---

[7]For simplicity, the time index is omitted.

[8]Superscripts are used to indicate the number of a source-encoder parameter and its corresponding output quantizer bit-vector.

[9]The LPC-coefficients (linear predictive coding) describe the spectral shape of a source vector.

not the mean-squared error of the reconstructed multimedia signal itself.[10] The reason is that the source codecs are complex algorithms that are hard to describe analytically. Hence, it is unknown how to (and probably impossible to) optimally estimate the output signal $\tilde{u}$ directly from the channel outputs $\tilde{z}$. The use of the mean-squared error in the parameter domain leads to algorithms (e.g., [Goertz (1999)]) with acceptable complexity that perform well, although they are inherently suboptimal. The block length N of the parameter vectors is determined by the source codec (e.g., $N = 10$, if the LPC coefficients of a narrow-band speech signal are quantized).

2.6.2 *Separation of Source and Channel Coding*

The foundation for the structure of most of today's transmission systems is the separation principle of information theory [McEliece (2002); Cover and Thomas (1991); Gallager (1968)]: it guarantees that—possibly but not necessarily by use of an infinitely large block length N—one may replace the encoder and the decoder mappings in Figure 2.1 by cascades of mappings with binary interfaces, at which independent and uniformly distributed bits are exchanged—without any loss in performance compared with the best possible direct mapping. Such a system is depicted in Figure 2.7. The

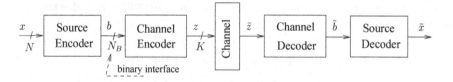

Fig. 2.7   Separation principle.

basic notion is to apply channel coding to achieve error-free bits at a bit rate that equals the channel capacity. A source code is applied to reduce the number of bits required to represent the source signal to the amount that can be transmitted error-free over the channel. Clearly, this reduction is only possible at the price of some distortion that is imposed on the source signal, if the latter has continuous amplitudes. It should be noticed that

---

[10]Even if it was possible to minimize the mean squared reconstruction error of the multimedia signal, this would still not lead to the best possible quality in terms of human perception, due to auditory masking, cognitive effects of human hearing, psycho-visual effects, and others which could be (but are not fully) exploited to design better coding algorithms. The modeling of these effects lies far beyond of the scope of this work.

the problem stated in Section 2.5.2, with its simple solution depicted in Figure 2.3, could also be solved by a separation into source and channel coding—a bad choice in this case, as the price is infinite block length and complexity. Nevertheless, with some modifications discussed below, source and channel coding are separate processing steps in all practical systems. The reason is that the separation allows to construct partial mappings, e.g., channel encoders/decoders, that, compared with asymptotic results, sometimes may have only moderate performances,[11] but their complexities are tolerable.

If the separation principle is applied in a system with limited block lengths, the following problems occur:

(1) it is impossible to realize channel codes that have a residual bit error rate of zero,
(2) the source encoder output bits are not independent and uniformly distributed, i.e., the coded bits contain residual redundancies,
(3) the distortion at the source decoder output depends on which bit is in error.

While the first issue is the reason why joint source-channel decoding is needed, the last two aspects are the basis for any measures against the quality decrease that arises from residual bit errors. We will briefly discuss the three issues below.

2.6.2.1 *Channel Codes have a Non-Zero Residual Bit Error Rate*

As an example, let us consider a binary symmetric channel with a bit error probability of $p_e = 5/63 = 0.0794$. In this case the channel capacity (for details, see Appendix A) equals $C = 0.6$ bits per channel-use.

Now consider the following two channel codes [Lin and Costello (1983)]:

- (7,4)-Hamming code, code rate $R_c = 4/7 = 0.571$, can correct all 1-bit errors
- (63,36)-BCH code, rate $R_c = 36/63 = 0.571$, can correct all 5-bit errors and less.

---

[11]It should be mentioned that, in the "waterfall-region," turbo codes [Berrou and Glavieux (1996)] and the developments based thereon brought the performance of channel coding with fairly long block-lengths close to the information theoretical limits. But as all other channel codes, turbo codes cannot guarantee a bit error rate of zero after decoding; in fact, turbo codes show an error floor below a residual bit error rate of, e.g., 10^{-5}.

Since both code rates R_c are below the channel capacity $(R_c < C)$, it is, according to information theory, in principle possible to obtain error-free bits at the channel decoder output. As illustrated by Figure 2.8, this is not true in practice. The vertical lines represent individual bit positions:

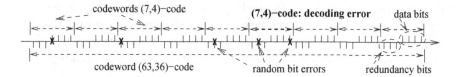

Fig. 2.8   Influence of the block length in channel coding.

the locations of the data bits are indicated by the vertical lines *above* the horizontal line and the locations of the redundancy bits are represented by the vertical lines *below* it. The codewords of the (7,4)-Hamming code are inserted as well as one codeword of the (63,36)-BCH code. Now we have randomly chosen locations of five bit errors, marked by crosses. In the 6-th codeword of the (7,4)-Hamming code we observe two bit errors, which means that the channel decoder will emit the wrong codeword, as only one bit error can be corrected. In contrast to that we have five bit errors in the (63,36)-BCH codeword which can be corrected. The problem with the short codewords of the Hamming code is that, due the randomness of the bit errors, sometimes there will be more than one error in a codeword and sometimes there will be no error at all. In the last case we don't need channel coding and in the first case we don't have enough redundancy bits to correct the errors. We can mitigate this problem by using codes with longer codewords, as we have shown by the example of the BCH-code, but as long as the block length is lower than infinity, the number of bit errors in a block will vary from one transmitted codeword to another, due to the randomness of the bit errors, so with some non-zero probability[12] there will be blocks with more errors than the code can correct. Only for infinitely large block length the relative number of bit errors in a block will converge against the bit error probability p_e so that we know how many errors will

---

[12]For a binary symmetric channel with bit error probability $p_e > 0$, the probability P_n of n bit errors at any locations within a K-bit codeword is given by

$$P_n = \binom{K}{n} \cdot p_e^n \cdot (1 - p_e)^{K-n} \quad \rightarrow \quad P_n > 0, \, n = 0, 1, ..., K \quad \text{for} \quad p_e > 0 \,.$$

have to be corrected in *each* received block and we are able to design the code appropriately.

As we are forced by the application and by complexity to use a limited block length, we could improve the reliability of the decoded channel codewords by reducing the code rate (more redundancy bits to improve the error-correction capabilities of the code). This means, however, that we can transmit less data bits, so we have to find a compromise between the reliability we (really) need and the amount of data we want to transmit. In any case we will not be able to obtain error-free bits after channel decoding.

2.6.2.2   *Source Encoder Output Bits Contain Redundancies*

In practice, vector quantization is not directly applied to the source samples due to complexity, but conceptually one may think of any multimedia source coding scheme as a vector quantizer with limited dimension.[13] Therefore, we will use two-dimensional vector quantization to show, that the quantizer output bit-vectors contain residual redundancies.

Figure 2.9 shows realizations of two successive samples of an uncorrelated source (upper two plots) and a correlated source (lower two plots). The crosses indicate the locations of the reproduction levels of a source optimized scalar quantizer (in two dimensions) and vector quantizers designed for the corresponding source signals in the plots. If we consider the upper left plot it is obvious that the four combined reproduction levels of the scalar quantizer in the corners are almost never used, because almost no realizations of the source signal vectors are located in these areas. Hence, the joint probability distribution of two successive quantizer bit-vectors is not uniform which means they contain redundancies. An improvement, indicated by the quantization SNRs printed in the plots, can be achieved by use of a source-optimized two-dimensional vector quantizer (upper right plot): now the locations of the vector quantizer code-vectors are better adapted to the two-dimensional PDF of the source signal. If we, however, think of four-dimensional source vectors, the PDF will be spherical but the two-dimensional code-vectors are not adapted to this, i.e., there will be combinations of successive two-dimensional code-vectors that are rarely used, so again there is redundancy left in the quantizer bit-vectors, due to their non-uniform probability distribution. The same argumentation is even more true for the correlated source signal in the lower two plots.

---

[13]Asymptotically, a vector quantizer achieves the rate-distortion function [Jayant and Noll (1984)].

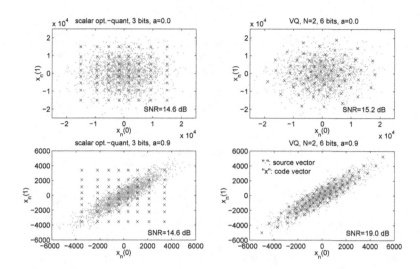

Fig. 2.9   Scalar and vector quantization of an uncorrelated and a correlated source.

Another way to show that there is always redundancy left in fixed-rate source-optimized quantizer outputs is based on the fact, that each quantizer level contributes the same amount of distortion at high rates [Jayant and Noll (1984)]. As the quantizer intervals (Voronoi regions) have different sizes (due to the adaptation of the quantizer to the PDF of the source signal), the larger intervals must be used less frequently so that their contribution to the total distortion equals that of the smaller intervals.

Over all, this means that as long as the vector dimension is not infinitely large, the quantizer levels contain residual redundancies and the coding system is not perfect in the sense of information theory. Moreover, this imperfectness is known a-priori at the receiver and can be exploited to improve the system.

2.6.2.3   *It Matters, which Bit is in Error*

Figure 2.10 shows the code-vectors of a two-dimensional three-bit vector quantizer and the associated bit-vectors. The lines connect each pair of the code-vectors whose bit-vectors differ in exactly one bit position; the leftmost bit position in the leftmost plot, the rightmost bit position in the rightmost plot. The length of the lines indicate the distortion that occurs when, due to a bit error during the transmission, the code-vectors are permuted at

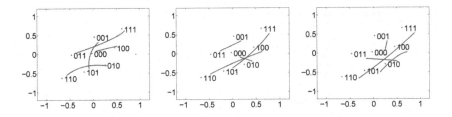

Fig. 2.10   Bad bit mapping for a vector quantizer.

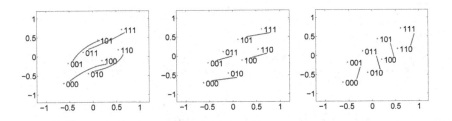

Fig. 2.11   Optimal bit mapping for a vector quantizer.

the decoder output. The average length of the lines is a measure for the distortion at the decoder output due to single bit errors.

Figure 2.11 shows the same information as Figure 2.10, but for a different bit mapping. It is obvious that the average distortion due to bit errors is lower for the (optimized) bit mapping in Figure 2.11, but it is also clear that it makes a strong difference for the distortion which bit (the leftmost or the rightmost) is in error. Thus, if a for a vector quantizer with limited block length the bit mapping is optimized so that the distortion due to bit errors is minimum, the optimal bit mapping will have more significant and less significant bits.

2.6.3   *Approaches to Joint Source-Channel Decoding*

2.6.3.1   *Unequal Error Protection and Error Concealment*

Figure 2.12 shows some system modifications that account for the drawbacks of the separation principle stated above. Implicitly, a channel with discrete input alphabet (e.g., by application of a digital modulation scheme) is assumed, because the channel codeword z—a bit-vector—is directly fed into the channel. As the output bits of the source encoder are not equally

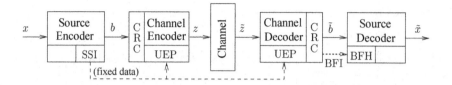

Fig. 2.12   Unequal error protection and error concealment.

sensitive to bit errors, some of the data bits cause stronger quality degradations than others if they are in error but are used for decoding anyhow. Source significance information (SSI) can be derived from that and be used by the channel-coding scheme to apply stronger error protection to the more sensitive bits, e.g., by puncturing of those output bits of a low-rate channel code that protect the less sensitive bits. This scheme is called unequal error protection (UEP) [Masnick and Wolf (1967)].

The residual redundancies in the source-encoder output bits can be exploited to conceal bit-errors (bad frame handling (BFH)), e.g., by repetition of the "old bits" from the previous block. Clearly, this only makes sense if the bits are correlated in time. For the initiation of the error concealment an error detection is required, which is usually realized by an error-detecting channel code (cyclic redundancy check (CRC)). If errors are detected this is communicated to the source decoder by the bad frame indication flag (BFI). The strongly sensitive bits of the source encoder are CRC-encoded (class-1-bits) and afterwards the CRC codeword is channel encoded together with all other remaining data bits. Thus, in combination with UEP, the bits of the source encoder are divided into several classes, which are protected by individually adjustable amounts of channel-code redundancy. Fortunately, the more sensitive bits are typically the stronger correlated ones at the same time, i.e., unequal error protection and error concealment can be combined very efficiently. Hence, both schemes are frequently used, e.g., in every mobile radio standard.

2.6.3.2   *Source-Controlled Channel Decoding*

The basic scheme is depicted in Figure 2.13. The idea is to exploit the residual redundancies—more systematically than above—as a-priori information in the decision procedure for the data bits in channel decoding to reduce the bit error rate. This concept can be implemented very efficiently for the decoding of binary convolutional codes; moreover, the idea can also be

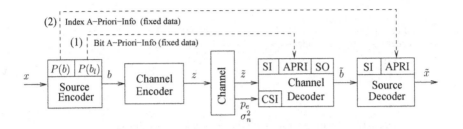

Fig. 2.13   (1) Source-controlled channel decoding and (2) estimation-based source decoding.

extended to soft-output decoders (**APRI-SOVA**, [Hagenauer (1995)]) which provide reliability information—and not only hard decisions—for the data bits. These reliabilities can be exploited by estimation-based source decoders described below. For a correct weighting of the a-priori information and the soft channel-outputs, a channel state information (CSI) must be available (e.g., the noise variance). A recent extension of this approach and applications to low-rate speech coding are stated in [Alajaji *et al.* (1996); Fazel and Fuja (2003)].

For brevity of notation, only the unconditional probability distributions $P(b_l)$ of the data bits are used in Figure 2.13, but the decoding algorithm can be easily extended to probability distributions that are conditioned on previously received channel outputs, possibly by use of a Markov-model for the time-dependencies of the bits.

A drawback is that the redundancies are usually exploited on bit-basis[14] (because mostly binary channel codes are used), although the source encoder often produces correlated *bit-vectors* that consist of several bits. Thus, a large part of the redundancies is removed if only the marginal distributions of the individual bits are used as a-priori information.

2.6.3.3   *Estimation-Based Source Decoding*

In conventional systems, simple procedures, e.g., quantizer table-look-ups, are frequently applied for source decoding. If, however, source encoding is

---

[14]In [Heinen *et al.* (1998)] a channel-decoding scheme for binary convolutional codes is given that is able to optimally exploit index-based a-priori information, if all the bits of an index are adjacently mapped into the input of the channel encoder. The method may, however, not be applicable because usually bit-reordering/interleaving is employed at the channel-encoder input to account for different sensitivities of the bits or to improve the distance properties of the concatenated code.

not perfect and residual redundancies—e.g., correlations in time—are left in the data bits, this a-priori knowledge can be exploited for better source decoding by performing (optimal) estimations of the source-encoder input [Phamdo and Farvardin (1994); Fingscheidt and Vary (1997); Skoglund (1999); Sayood and Borkenhagen (1991); Miller and Park (1998)]; such a system is also depicted in Figure 2.13.

Conceptually, the idea is similar to the one described in the previous section, but now it is possible to exploit the full correlations within the bit-vectors (i.e., on *index*-basis); details will be given in Chapter 3. The soft-in/soft-out (SISO) channel decoder acts as a device that improves the reliability of the virtual channel "seen" by the source decoder. As in the previous section, bit-based a-priori information may be used to aid channel decoding, but it is not clear, how the a-priori information shall be appropriately divided between the channel decoder and the source decoder.

2.6.3.4 *Iterative Source-Channel Decoding*

In [Goertz (2001b); Perkert *et al.* (2001)] iterative source-channel decoding is introduced. Although the sketch of the system in Figure 2.14 is similar to the systems in Figure 2.13, the theoretical concept is different: the

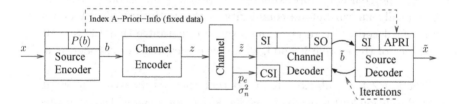

Fig. 2.14   Iterative source-channel decoding.

transmitter is interpreted as a serially concatenated channel-coding scheme; the constituent "codes" are the implicit residual redundancies within the source encoder output bit-vectors and the explicit redundancy of the channel code. This calls for the use of the turbo principle [Berrou and Glavieux (1996); Hagenauer *et al.* (1996)] for decoding. As in all iterative decoding schemes, decoders for both constituent "codes" must be available that are able to exchange extrinsic information on the data bits within the iterations. While such decoders for convolutional channel codes are well-known from literature [Hagenauer *et al.* (1996)], they can also be formulated on the basis of estimation-based decoding for the source-coding-part of the system, which we will show in Chapter 3 along with the fact that the iterative decoding scheme is a well-defined approximation of the prohibitively complex optimal joint decoder. Moreover, we will introduce a procedure for the optimization of the quantizer bit mappings for the iterative decoding process, which leads to further strong gains in transmission power. Among the realizable decoding schemes, iterative source-channel decoding works best but at the same time it is the most complex approach.

Iterative source-channel decoding can also be applied to the serial concatenation of variable-length codes (VLC) and convolutional channel codes. This topic was treated, e.g., in [Bauer (2002); Park and Miller (2000)]. In our work, however, we don't consider VLCs as the focus lies on the low-delay transmission of continuous sources with fixed block lengths as, e.g., in conversational applications such as telephony in mobile radio.

2.6.4 *Approaches to Joint Source-Channel Encoding*

In Section 2.6.3, several practical approaches to partially rejoin source and channel *de*coding were briefly described. In all schemes the imperfectness of real-world source and channel encoding algorithms is exploited for better decoding. The goal is to approximate the optimal joint decoder, which is algorithmically known but its complexity is prohibitively high. A close-to-optimum decoder design gives, however, no guarantee that the overall performance of the system is good, as we have accepted the *en*coder as it is.

In Section 2.4, we already discussed that there is no straightforward analytical solution for the optimal encoder design; hence we will concentrate on three special problems that imply some constraints that give room for system improvements.

In Chapter 4, we will show that channel-optimized vector quantization (COVQ) requires to store several codebooks, one for each value of the bit-error probability, if COVQ is used on time-varying channels. We will introduce a new algorithm called channel-adaptive scaled vector quantization (CASVQ) that approximates COVQ by only one scaled reference codebook; the gains of COVQ over source-optimized VQ are widely retained by the scheme.

In Chapter 5, we will consider packet based multimedia transmission. In particular, we will discuss, how to optimize multiple index assignments for vector quantizers in such a way, that the performance in the source-

SNR-sense is maximized. This involves the optimal multiple description (MD) decoder but also an optimization algorithm for index assignment at the transmitter. The algorithm allows to use standardized multimedia source codecs in packet based transmission without affecting the parameter quantizers that are used in the codecs. Hence, the scheme is compatible with any source coding standard it is applied to.

In Chapter 6, we will introduce a new algorithm that quantizes a continuous-valued source signal and transmits it by a source-adaptive digital modulation scheme. The modulation signal points are shifted at the transmitter side, dependent on the *current unquantized* value at the input. Each shift is individually adapted to the current unquantized input source signal such that the influence of the channel noise on the receiver output signal is minimized. A conventional receiver is used, i.e., the new algorithm requires only modifications at the transmitter, which makes it attractive for broadcast scenarios, but also for the down-link in a mobile radio network. In Chapter 7, we will discuss a variation of source-adaptive modulation, in which we use a binary or quarternary transmission, but the power of each quantizer bit is individually adjusted such that the expected distortion at the source decoder output is minimized.

Chapter 3

Joint Source-Channel Decoding*

3.1 Introduction and System Model

In this chapter, we will use the system model in Figure 3.1. It is a general-
ization of the system in Figure 2.14 that has been briefly discussed in the
introductory overview. A set of mutually independent but auto-correlated

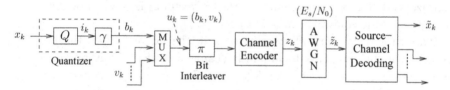

Fig. 3.1 System model (reprinted from [Hagenauer and Goertz (2003)] with permission,
©2003 IEEE).

input source signals is transmitted at each time index k. This is a realistic
model for the coded transmission of, e.g., speech, audio, or image signals,
where the actual source signal is decomposed into parameters—the source
signals in our system model—which are independently quantized, jointly
channel encoded, and transmitted over a noisy channel. As it is required in
many practical cases due to delay constraints, we assume that decoding at
the receiver must be carried out instantaneously after a channel codeword

---

*Parts of this chapter are based on (i) "A Generalized Framework for Iterative Source-
Channel Decoding" by N. Goertz, which appeared in *Annals of Telecommunications*,
pp. 435–446, July/August 2001, ©2001 Annals of Telecommunications, and (ii) "The
Turbo Principle in Joint Source-Channel Coding" by J. Hagenauer and N. Goertz, which
appeared in *Proceedings Information Theory Workshop*, pp. 275–278, Paris/France,
April 2003, ©2003 IEEE.

has been received. Moreover, the auto-correlations of the source-codec parameters (e.g., the energy of a speech signal) can frequently be accurately modeled by first-order Markov-processes; this will be detailed in what follows.

To simplify the notation, we will only consider one of the inputs, the samples (codec parameters) x_k in Figure 3.1, which are quantized at each time k with N_B bits by the indices i_k. For the transmission, the indices i_k (the row-numbers in the codebook in which the quantizer code-vectors/reproduction levels y_{i_k} are located) are mapped to the bit-vectors

$$b_k \doteq \{b_{k,1}, ..., b_{k,n}, ..., b_{k,N_B}\} \in \mathcal{B} \doteq \{0,1\}^{N_B} \qquad (3.1)$$

by

$$b_k = \gamma(i_k) , \quad i_k \in \{0, 1, ..., 2^{N_B} - 1\} , \qquad (3.2)$$

where γ is the deterministic mapping function and $\mathcal{B}$ is the set of all possible N_B-bit vectors. The inverse of (3.2), the mapping from a bit-vector b_k to the corresponding quantizer index i_k, is given by $i_k = \gamma^{-1}(b_k)$. The quantizer code-vector corresponding to the bit-vector b_k is denoted by $y(b_k)$ to abbreviate the equivalent notation $y_{\gamma^{-1}(b_k)}$.

Placed together[1] with all parallel data in the bit-vector u_k, the bit-vectors b_k are bit-interleaved and jointly channel-encoded; the K-bit channel codeword $z_k = \{z_{k,l}, l = 1, ..., K\}$, $z_k \in \mathcal{Z} \doteq \{0,1\}^K$, is transmitted over an AWGN-channel. Since we assume coherently detected binary modulation (binary phase-shift keying), the conditional PDF[2] of the channel output value $\tilde{z}_{k,l}$, given that the code bit $z_{k,l} \in \{0,1\}$ has been transmitted, is given by

$$p_c(\tilde{z}_{k,l}|z_{k,l}) = \frac{1}{\sqrt{2\pi}\sigma_n} e^{-\frac{1}{2\sigma_n^2}(\tilde{z}_{k,l} - (1 - 2 \cdot z_{k,l}))^2} , \qquad (3.3)$$

with the variance[3] $\sigma_n^2 = \frac{N_0}{2E_s}$. E_s is the energy that is used to transmit each channel-code bit and $N_0/2$ is the power spectral density of the channel noise. The joint conditional PDF $p_c(\tilde{z}_k|z_k)$ for a channel word $\tilde{z}_k \in \tilde{\mathcal{Z}} \doteq \mathbb{R}^K$ to be received, given that the codeword $z_k \in \mathcal{Z}$ is transmitted, is the product of

---

[1] Without loss of generality, we assume that the bits of the bit-vector b_k are located at the first positions in the bit-vector u_k. This simplifies the notation below.

[2] For brevity we will only use the realizations of the random variables—denoted by small letters—in all formulas, as long as there is no risk of confusion. The random variables are denoted by capital letters; they are introduced if required.

[3] We use the power-normalized channel model, in which the power of the input samples is one. For details, see Appendix A.3.3.4.

(3.3) over all code-bits, since the channel noise is statistically independent, i.e.,

$$p_c(\tilde{z}_k | z_k) = \prod_{l=1}^{K} p_c(\tilde{z}_{k,l} | z_{k,l}) \,. \tag{3.4}$$

Source Model

As the source samples x_k are assumed to be correlated, adjacent quantizer indices i_{k-1}, i_k show dependencies. They are modeled by first-order stationary Markov-processes, which are described by the index transition probabilities $\mathrm{P}(i_k | i_{k-1})$ or—due to (3.2) completely equivalent—by the bit-vector transition probabilities $\mathrm{P}(b_k | b_{k-1})$. Figure 3.2 shows the correspond-

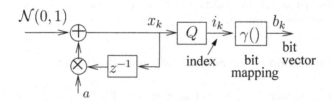

Fig. 3.2   First-order Gauss-Markov source model and quantization.

ing source model, in which, as an example, an optimal scalar quantizer is used (Lloyd-Max quantizer [Jayant and Noll (1984)]). Figure 3.3 shows the conditional probability distribution of successive quantizer indices[4] for a strongly correlated ($a = 0.9$) source signal, and Figure 3.4 shows the allocation of the quantizer levels y_i to their indices and the unconditional probability distribution of the quantizer levels. It is obvious that the indices are strongly correlated in time, as, for instance, the index succeeding $i_{k-1} = 0$ will take on values in the range $i_k = 0, 1, ..., 9$ almost with probability one, but the indices $i_k = 10, 11, ..., 31$ will (almost) never be used. Figure 3.5 shows the conditional probability distributions of successive quantizer indices for a weakly correlated ($a = 0.5$) source signal.

---

[4]We show the conditional *index* probabilities $\mathrm{P}(i_k | i_{k-1})$ and not the conditional bit-vector probabilities because we use a scalar optimal quantizer with quantizer levels that are sorted in the codebook (see Figure (3.4)). Thus, the correlation of adjacent quantizer outputs can easily be seen in Figure 3.3.

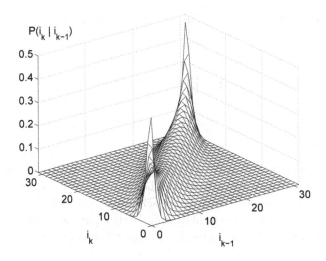

Fig. 3.3 Conditional probability distribution of successive 5-bit quantizer indices for $a = 0.9$ (strong low-pass correlation).

3.2   Near Optimum Joint Source-Channel Decoding

The goal is to minimize the mean-squared error (MSE) of the decoder output $\tilde{x}_k$, i.e., we want to perform joint source-channel *decoding* (JSCD) for the fixed transmitter described above. Hence, we minimize the conditional expected MSE

$$D_D(\tilde{\boldsymbol{z}}_k) \doteq \mathrm{E}_{X_k | \tilde{\boldsymbol{z}}_k}\left\{ \left\| X_k - \tilde{x}_k \right\|_2^2 \, \Big| \, \tilde{\boldsymbol{z}}_k \right\} \tag{3.5}$$

where $\tilde{\boldsymbol{z}}_k \doteq \{\tilde{z}_0, \tilde{z}_1, ..., \tilde{z}_k\}$ is the set of all received channel-output vectors up to the current time k. The whole past of the channel outputs is included in (3.5) because we want to exploit the correlation of the source signal for decoding. Although we may have a set of independent parallel input signals in Figure 3.1 that are jointly channel encoded and transmitted, we can separately minimize (3.5) for each signal by a proper choice of the decoder output $\tilde{x}_k$, because the mean-squared error $\left\| x_k - \tilde{x}_k \right\|_2^2$ is non-negative for any x_k, $\tilde{x}_k$ (if we use the average MSE of all parallel signals as a performance measure, we cannot compensate for a positive MSE of one signal by a negative "error" of another). We can expand (3.5) as follows:

$$D_D(\tilde{\boldsymbol{z}}_k) = \mathrm{E}_{X_k | \tilde{\boldsymbol{z}}_k}\left\{ \frac{1}{N} \sum_{i=0}^{N-1} (X_{k,i} - \tilde{x}_{k,i})^2 \, \Big| \, \tilde{\boldsymbol{z}}_k \right\}. \tag{3.6}$$

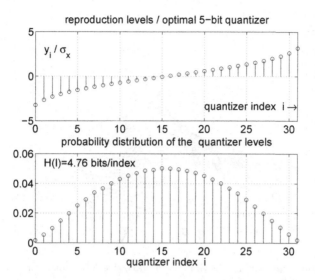

Fig. 3.4 Optimal 5-bit scalar quantizer for a Gaussian source: index assignment, quantizer levels y_i, and their unconditional probability distribution.

To find the minimum of $D_D(\tilde{\boldsymbol{z}}_{\boldsymbol{k}})$ we take the partial derivatives with respect to the components $\tilde{x}_{k,j}$ of the decoder output vector:

$$\frac{\partial D_D(\tilde{\boldsymbol{z}}_{\boldsymbol{k}})}{\partial x_{k,j}} = -\frac{2}{N} \cdot \mathrm{E}_{X_k|\tilde{\boldsymbol{z}}_{\boldsymbol{k}}}\left\{(X_{k,j} - \tilde{x}_{k,j}) \,\big|\, \tilde{\boldsymbol{z}}_{\boldsymbol{k}}\right\}, \quad j = 0, 1, ..., N-1 \,. \quad (3.7)$$

Set to zero, we obtain the minimum mean-square estimator $\tilde{x}_{k,j}^{\circledast}(\tilde{\boldsymbol{z}}_{\boldsymbol{k}}) = \mathrm{E}_{X_k|\tilde{\boldsymbol{z}}_{\boldsymbol{k}}}\left\{X_{k,j} \,\big|\, \tilde{\boldsymbol{z}}_{\boldsymbol{k}}\right\}$ as the optimal component decoder, which we may write in vector notation according to

$$\tilde{x}_k^{\circledast}(\tilde{\boldsymbol{z}}_{\boldsymbol{k}}) = \mathrm{E}_{X_k|\tilde{\boldsymbol{z}}_{\boldsymbol{k}}}\left\{X_k \,\big|\, \tilde{\boldsymbol{z}}_{\boldsymbol{k}}\right\}. \quad (3.8)$$

By expansion of (3.8), we find

$$\tilde{x}_k^{\circledast}(\tilde{\boldsymbol{z}}_{\boldsymbol{k}}) = \int_{\mathcal{X}} x_k \cdot p(x_k|\tilde{\boldsymbol{z}}_{\boldsymbol{k}})dx_k = \int_{\mathcal{X}} x_k \frac{p(\tilde{\boldsymbol{z}}_{\boldsymbol{k}}|x_k) \cdot p(x_k)}{p(\tilde{\boldsymbol{z}}_{\boldsymbol{k}})}dx_k \,; \quad (3.9)$$

the second equality follows from the Bayes-rule. Now we split the integration over all possible inputs $x_k \in \mathcal{X}$ into a sum of integrals over the "quantizer intervals" $\mathcal{X}_j$:

$$\tilde{x}_k^{\circledast}(\tilde{\boldsymbol{z}}_{\boldsymbol{k}}) = \sum_{i=0}^{2^{N_B}-1} \frac{1}{p(\tilde{\boldsymbol{z}}_{\boldsymbol{k}})} \int_{\mathcal{X}_i} x_k \cdot p(\tilde{\boldsymbol{z}}_{\boldsymbol{k}}|x_k) \cdot p(x_k)dx_k \,. \quad (3.10)$$

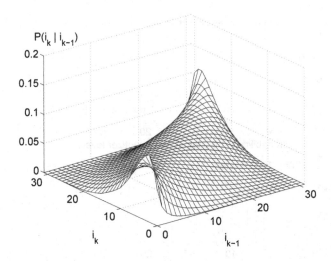

Fig. 3.5   Conditional probability distribution of successive 5-bit quantizer indices for $a = 0.5$ (weak low-pass correlation).

To simplify (3.10), we aim to replace $p(\tilde{\boldsymbol{z}}_k|x_k)$ by $p(\tilde{\boldsymbol{z}}_k|i_k = Q(x_k))$, i.e., we want to represent the source vector x_k by its quantizer index i_k. If the source was memoryless, this would mean no approximation. In our case, however, the source vector x_k carries information about $\tilde{z}_k$ but, due to the source correlation, also about $\tilde{z}_{k-1}, \tilde{z}_{k-2}, \dots$. Hence, the condition x_k influences the "probabilities" of all received channel output vectors contained in $\tilde{\boldsymbol{z}}_k = \{\tilde{z}_0, \tilde{z}_1, \dots, \tilde{z}_{k-1}, \tilde{z}_k\}$. If we replace the source vector x_k by its quantizer index, we discard the information about the exact location of the source vector in the quantizer interval and, hence, we lose correlation-based information about the location of the source vectors $x_{k-1}, x_{k-2}, \dots$, which, after being quantized, directly lead to the channel input vectors $z_{k-1}, z_{k-2}, \dots$. The loss of information is small ("near optimum decoding"), if a high-rate quantizer is used, which we will assume in what follows. The optimal decoder (3.8) would require an increasing amount of memory with growing time index k, making its implementation infeasible.[5] Using the approximation, we obtain from (3.10):

$$\tilde{x}_k^{\circledast}(\tilde{\boldsymbol{z}}_k) = \sum_{i_k=0}^{2^{N_B}-1} \frac{1}{p(\tilde{\boldsymbol{z}}_k)} \cdot p(\tilde{\boldsymbol{z}}_k|i_k) \int_{\mathcal{X}_{i_k}} x_k \cdot p(x_k) dx_k \ . \qquad (3.11)$$

[5] The author would like to thank Mikael Skoglund from Royal Institute of Technology (KTH), Stockholm, for pointing out subtle details of that in a personal conversation.

With the centroid (optimal reproducer vector, see Section 2.42)

$$y_{i_k} = \frac{1}{\mathrm{P}(i_k)} \int_{\mathcal{X}_{i_k}} x_k \cdot p(x_k) dx_k \qquad (3.12)$$

of a vector quantizer for the quantizer decision region $\mathcal{X}_{i_k}$, we find the decoder

$$\tilde{x}_k^{\circledast}(\tilde{z}_k) = \sum_{i_k=0}^{2^{N_B}-1} y_{i_k} \cdot \mathrm{P}(i_k \mid \tilde{z}_k) = \sum_{b_k \in \mathcal{B}} y(b_k) \cdot \mathrm{P}(b_k \mid \tilde{z}_k) , \qquad (3.13)$$

as $\frac{p(\tilde{z}_k|i_k)\mathrm{P}(i_k)}{p(\tilde{z}_k)} = \mathrm{P}(i_k|\tilde{z}_k)$; for the right equality we used the equivalent bit-vector notation.

Hence, the main problem of joint source-channel decoding (3.13) is how to compute the *bit-vector a-posteriori probabilities* $\mathrm{P}(b_k \mid \tilde{z}_k)$. By use of the Bayes-rule we can show that

$$\mathrm{P}(b_k \mid \tilde{z}_k) = \frac{p(b_k, \tilde{z}_k, \tilde{z}_{k-1})}{p(\tilde{z}_k)} = \underbrace{\frac{p(\tilde{z}_{k-1})}{p(\tilde{z}_k)}}_{\doteq A_k} \cdot \underbrace{p(\tilde{z}_k \mid b_k, \tilde{z}_{k-1})}_{\text{channel term}} \cdot \underbrace{\mathrm{P}(b_k \mid \tilde{z}_{k-1})}_{\text{a-priori term}} ,$$

$$(3.14)$$

where $\mathrm{P}(b_k \mid \tilde{z}_{k-1})$ is the *bit-vector a-priori probability* at time k and A_k is a factor that ensures that the left-hand side of (3.14) sums up to one over all $b_k \in \mathcal{B}$. Thus, A_k can be computed from this condition, and the PDFs $p(\tilde{z}_{k-1})$, $p(\tilde{z}_k)$ do not need to be known.

The a-priori term in (3.14) can be computed as follows:

$$\mathrm{P}(b_k \mid \tilde{z}_{k-1}) = \sum_{b_{k-1} \in \mathcal{B}} \mathrm{P}(b_k, b_{k-1} \mid \tilde{z}_{k-1})$$

$$= \sum_{b_{k-1} \in \mathcal{B}} \frac{\mathrm{P}(b_k \mid b_{k-1}, \tilde{z}_{k-1}) \cdot p(b_{k-1}, \tilde{z}_{k-1})}{p(\tilde{z}_{k-1})} . \qquad (3.15)$$

As the probability for the bit-vector at time k is not influenced by the channel outputs received in the past if the bit-vector at time $k-1$ is known, we have $\mathrm{P}(b_k \mid b_{k-1}, \tilde{z}_{k-1}) = \mathrm{P}(b_k \mid b_{k-1})$. Hence, the *bit-vector* a-priori probabilities are given by

$$\mathrm{P}(b_k \mid \tilde{z}_{k-1}) = \sum_{b_{k-1} \in \mathcal{B}} \underbrace{\mathrm{P}(b_k \mid b_{k-1})}_{\text{Markov-model}} \cdot \underbrace{\mathrm{P}(b_{k-1} \mid \tilde{z}_{k-1})}_{\text{old APPs}} , \qquad (3.16)$$

i.e., they are computed from the "old" *bit-vector* APPs (3.14) at time $k-1$ and from the transition probabilities of the Markov model for the quantizer indices. The information carried by the currently received channel-word

$\tilde{z}_k$ is not used in (3.16). For initialization at time $k=0$ the unconditional probability distribution of the bit-vectors can be used.

The "channel term" in (3.14) is much harder to compute than the a-priori term, if several bit-vectors are commonly channel encoded. For convenience of notation, we define the bit-vector v_k that contains all the other bits that are jointly channel encoded together with the bits in b_k, but v_k does not contain the bit-vector b_k. As defined in Figure 3.1, the bit-vector u_k contains all bits that are jointly channel-encoded, i.e.,

$$u_k = (b_k, v_k) \,. \tag{3.17}$$

In a practical application, v_k may typically consist of several hundreds (or even thousands) of bits. The set of all possible bit-combinations for v_k is denoted by $\mathcal{V}$. With these definitions we can rewrite the channel-term in (3.14) as follows:

$$p(\tilde{z}_k \mid b_k, \tilde{\boldsymbol{z}}_{k-1}) = \sum_{v_k \in \mathcal{V}} \frac{p(\tilde{z}_k, \overbrace{b_k, v_k}^{= u_k}, \tilde{\boldsymbol{z}}_{k-1})}{p(b_k, \tilde{\boldsymbol{z}}_{k-1})} \tag{3.18}$$

$$= \sum_{v_k \in \mathcal{V}} \frac{p(\tilde{z}_k \mid (b_k, v_k), \tilde{\boldsymbol{z}}_{k-1}) \cdot p((b_k, v_k), \tilde{\boldsymbol{z}}_{k-1})}{p(b_k, \tilde{\boldsymbol{z}}_{k-1})} \tag{3.19}$$

$$= \sum_{v_k \in \mathcal{V}} p(\tilde{z}_k \mid (b_k, v_k), \tilde{\boldsymbol{z}}_{k-1}) \cdot \mathrm{P}(v_k \mid b_k, \tilde{\boldsymbol{z}}_{k-1}) \,. \tag{3.20}$$

As the channel is assumed to be memoryless, the past of the received channel words contained in $\tilde{\boldsymbol{z}}_{k-1}$ does not provide additional information about $\tilde{z}_k$ if (b_k, v_k) is given, i.e., $p(\tilde{z}_k \mid (b_k, v_k), \tilde{\boldsymbol{z}}_{k-1}) = p(\tilde{z}_k \mid (b_k, v_k))$. Moreover, the channel input deterministically depends on the input bit-vector $u_k = (b_k, v_k)$, i.e.,

$$p(\tilde{z}_k \mid (b_k, v_k), \tilde{\boldsymbol{z}}_{k-1}) = p_c(\tilde{z}_k \mid z(b_k, v_k)) \,, \tag{3.21}$$

where $p_c(\cdot)$ denotes the channel PDF (3.4) and $z(b_k, v_k))$ is the channel-input word that results from the channel encoder input bit-vector $u_k = (b_k, v_k)$.

The rightmost term under the sum in (3.20) is the a-priori information for the "other" bits in the bit-vector v_k. As the bits in v_k and b_k are independent (this was assumed in the system model) we have

$$\mathrm{P}(v_k \mid b_k, \tilde{\boldsymbol{z}}_{k-1}) = \mathrm{P}(v_k \mid \tilde{\boldsymbol{z}}_{k-1}) \,. \tag{3.22}$$

Hence the computation of the a-posteriori probabilities in (3.13) can be summarized as follows:

$$\underbrace{\mathrm{P}(b_k \mid \tilde{\boldsymbol{z}}_k)}_{\text{current APP}} = A_k \cdot p(\tilde{z}_k \mid b_k, \tilde{\boldsymbol{z}}_{k-1}) \cdot \underbrace{\mathrm{P}(b_k \mid \tilde{\boldsymbol{z}}_{k-1})}_{\text{a-priori info}} \tag{3.23}$$

$$\frac{1}{A_k} \doteq \frac{p(\tilde{z}_k)}{p(\tilde{z}_{k-1})} = \sum_{b \in \mathcal{B}} p(\tilde{z}_k \mid b_k, \tilde{z}_{k-1}) \cdot P(b_k \mid \tilde{z}_{k-1}) \tag{3.24}$$

$$\underbrace{P(b_k \mid \tilde{z}_{k-1})}_{\text{a-priori info}} = \sum_{b_{k-1} \in \mathcal{B}} \underbrace{P(b_k \mid b_{k-1})}_{\text{Markov-model}} \cdot \underbrace{P(b_{k-1} \mid \tilde{z}_{k-1})}_{\text{old APPs}}, \tag{3.25}$$

$$p(\tilde{z}_k \mid b_k, \tilde{z}_{k-1}) = \sum_{v_k \in \mathcal{V}} \underbrace{p_c(\tilde{z}_k \mid z(b_k, v_k))}_{\text{channel PDF}} \cdot \underbrace{P(v_k \mid \tilde{z}_{k-1})}_{\text{a-priori info for other bits}} \tag{3.26}$$

Equation (3.24) is an alternative way to compute the normalization factor $A_k \doteq p(\tilde{z}_{k-1})/p(\tilde{z}_k)$, which ensures that the result of (3.23) is a true probability that sums up to one. Equation (3.25) is a copy of (3.16); (3.26) is given by (3.20), (3.21), (3.22).

In practice, this scheme to compute the bit-vector APPs in (3.13) can only be used, if the number of bits in the bit-vectors b_k and v_k is small. While this may be true for b_k also in practical cases[6] it is mostly not true for v_k, as this bit-vector consists of all bits of all other source-codec parameters (typically a few hundred in speech coding). Hence, the computation of (3.26) is usually infeasible due to complexity.

3.2.1 *Specialization and Generalization*

One special case of the joint source-channel decoding algorithm is worth noting: if $u_k = b_k$, i.e., only one signal channel exists in Figure 3.1, the channel term given by (3.26) simplifies to

$$p(\tilde{z}_k \mid b_k, \tilde{z}_{k-1}) = p(\tilde{z}_k \mid b_k) = p_c(\tilde{z}_k \mid z(b_k)) \tag{3.27}$$

because the channel is memoryless and $\tilde{z}_{k-1}$ does not provide further information on $\tilde{z}_k$ if b_k is given. Thus, the joint source-channel decoding algorithm for a signal that is separately quantized and channel encoded is given by

$$\underbrace{P(b_k \mid \tilde{z}_k)}_{\text{current APP}} = A_k \cdot p_c(\tilde{z}_k \mid z(b_k)) \cdot \underbrace{P(b_k \mid \tilde{z}_{k-1})}_{\text{a-priori info}} \tag{3.28}$$

$$\frac{1}{A_k} = \sum_{b \in \mathcal{B}} p_c(\tilde{z}_k \mid z(b_k)) \cdot P(b_k \mid \tilde{z}_{k-1}) \tag{3.29}$$

---

[6] In speech coding, the quantizer indices for source-codec parameters typically have less than 10 bits.

$$\underbrace{\mathrm{P}\big(b_k \mid \tilde{z}_{k-1}\big)}_{\text{a-priori info}} = \sum_{b_{k-1} \in \mathcal{B}} \underbrace{\mathrm{P}(b_k | b_{k-1})}_{\text{Markov-model}} \cdot \underbrace{\mathrm{P}(b_{k-1} | \tilde{z}_{k-1})}_{\text{old APPs}}, \qquad (3.30)$$

Although this solution, which was called "channel-coded optimal estimation[7] (CCOE)" and was first stated in [Goertz (1998)], seems to be more specialized, a completely general solution can be derived from it: the bit-vector b_k may be interpreted as a "super source vector" that again consists of many quantizer indices that are coded in parallel by the channel code. The resulting solution does *not* require the parallel signal channels to be independent, i.e., the solution is rather general. We will, however, not state the formulas as their derivation is straightforward and the notation necessarily becomes quite involved.

When no channel coding is used in the system, the channel coding term $p_c(\tilde{z}_k \mid z(b_k))$ in (3.28) is only used for the data bits as there are no parity bits, i.e., $z(b_k) = b_k$. The algorithm resulting from this simplification is called "optimal estimation (OE)" in our context. This is to point out that the decoder output $\tilde{x}_k$ is estimated by (3.13) using all knowledge that is available at the decoder. The scheme was introduced in [Phamdo and Farvardin (1994); Fingscheidt and Vary (1997)].

3.3   Iterative Source-Channel Decoding (ISCD)

3.3.1   *Principle and Derivation*

Even if the number of jointly channel-encoded data bits (size of u_k) is only moderate, near optimum decoding as stated above is practically infeasible due to the tremendous complexity in the computation of $p\big(\tilde{z}_k \mid b_k, \tilde{z}_{k-1}\big)$ by (3.26). Therefore, a less complex way to compute an approximation was introduced in [Goertz and Heute (2000); Goertz (2000)] and is described in detail in [Goertz (2001b,a)].

We will briefly repeat the derivation: first, we re-write the left-hand side of (3.26),

$$p\big(\tilde{z}_k | b_k, \tilde{z}_{k-1}\big) = \frac{p\big(\tilde{z}_k, b_k, \tilde{z}_{k-1}\big)}{p\big(b_k, \tilde{z}_{k-1}\big)}, \qquad (3.31)$$

and then we replace the *bit-vector* probability densities by the products

---

[7]The name is used to point out that the source signal is estimated by the decoder using all knowledge that is available. Channel coding is included in the estimation process.

over the corresponding *bit* probability densities:

$$p(\tilde{z}_k|b_k, \tilde{z}_{k-1}) \approx \prod_{n=1}^{N_B} \frac{p(\tilde{z}_k, b_{k,n}, \tilde{z}_{k-1})}{p(b_{k,n}, \tilde{z}_{k-1})} = \prod_{n=1}^{N_B} \frac{P(b_{k,n} \mid \tilde{z}_k)}{P(b_{k,n} \mid \tilde{z}_{k-1})} \cdot \underbrace{\frac{p(\tilde{z}_k)}{p(\tilde{z}_{k-1})}}_{=1/A_k} .$$

(3.32)

If we insert (3.32) into (3.23) we obtain the approximation for the bit-vector APPs

$$P(b_k \mid \tilde{z}_k) \approx P(b_k \mid \tilde{z}_{k-1}) \cdot \prod_{n=1}^{N_B} \frac{P(b_{k,n} \mid \tilde{z}_k)}{P(b_{k,n} \mid \tilde{z}_{k-1})} .$$

(3.33)

The *bit* a-posteriori probabilities $P(b_{k,n}|\tilde{z}_k)$ can be efficiently computed by the symbol-by-symbol APP algorithm [Hagenauer *et al.* (1996)] for a *binary* convolutional channel code with a small number of states. Although the APP-algorithm only decodes the currently received channel word $\tilde{z}_k$, it still uses all the received channel words $\tilde{z}_k$ up to the current time k, because its bit-based a-priori information

$$P(b_{k,n} \mid \tilde{z}_{k-1}) = \sum_{b_k \in \mathcal{B}|b_{k,n}} P(b_k \mid \tilde{z}_{k-1})$$

(3.34)

(for a specific bit $b_{k,n} \in \{0,1\}$) is derived from the time-correlations (3.25) of the bit-vectors at the quantizer output.

We can interpret the fraction in (3.33) as the extrinsic information $P_e^{(C)}(b_{k,n})$ [Hagenauer *et al.* (1996); Berrou and Glavieux (1996)] that we get from the channel decoder and write

$$P(b_k \mid \tilde{z}_k) \approx P(b_k \mid \tilde{z}_{k-1}) \cdot \left[\prod_{n=1}^{N_B} P_e^{(C)}(b_{k,n}) \right] .$$

(3.35)

Note that we have introduced the superscript "(C)" to indicate that $P_e^{(C)}(b_{k,n})$ is the extrinsic information produced by the APP channel-decoding algorithm.

In principle, we now could compute the mean-square estimates $\tilde{x}_k$ for transmitted signals by (3.13) using the *bit-vector* APPs from (3.35), but the latter are only approximations of the optimal values, since the *bit* a-priori informations that were used for APP channel decoding did not contain the mutual dependencies (which we actually want to exploit for decoding) of the bits within the bit-vectors: they were removed by the summation in (3.34).

The idea to make up (in part) for this loss of information due to the bit-based information exchange between the constituent APP decoders is to perform several iterations of the constituent APP decoders; the basic notion is taken over from iterative decoding of turbo codes [Hagenauer *et al.* (1996)]. Hence, new *bit* APPs are computed from the intermediate results (3.35) by

$$P^{(S)}(b_{k,n} \mid \tilde{z}_k) = \sum_{b_k \in \mathcal{B}|b_{k,n}} P(b_k \mid \tilde{z}_k) \, . \qquad (3.36)$$

The superscript "(S)" was introduced, as (3.36) is computed after source decoding. Now, we can derive new *bit* extrinsic information from the source decoder by

$$P_e^{(S)}(b_{k,n}) \doteq P^{(S)}(b_{k,n} \mid \tilde{z}_k)/P_e^{(C)}(b_{k,n}) \, , \qquad (3.37)$$

where in the numerator the result of (3.36) is inserted. The extrinsic information from the last run of the channel decoder is removed by the denominator in (3.37), since we do not want to loop back information to the channel decoder that it has produced itself in the previous iteration. The extrinsic information computed by (3.37) is used as the new a-priori information for the second and further runs of the APP channel decoder.

Summary of iterative source-channel decoding (ISCD):

(1) At each time k, compute the initial *bit-vector* a-priori probabilities by (3.25).
(2) Use the results from Step 1 in (3.34) to compute the initial *bit* a-priori information for APP channel decoding.
(3) Perform APP channel decoding.
(4) Perform source decoding by inserting the extrinsic *bit* information from APP channel decoding into (3.35) to compute new (temporary) *bit-vector* APPs.
(5) If this is the last of a given number of iterations, proceed with Step 8, otherwise continue with Step 6.
(6) Use the *bit-vector* APPs of Step 4 in (3.36), (3.37) to compute extrinsic *bit* information from the source redundancies.
(7) Set the extrinsic *bit* information from Step 6 equal to the new *bit* a-priori information for the APP channel decoder in the next iteration; proceed with Step 3.
(8) Estimate the receiver output signals by (3.13) using the *bit-vector* APPs from Step 4.

The procedure described above is illustrated by Figure 3.6, where the dashed box marks the APP source decoder.

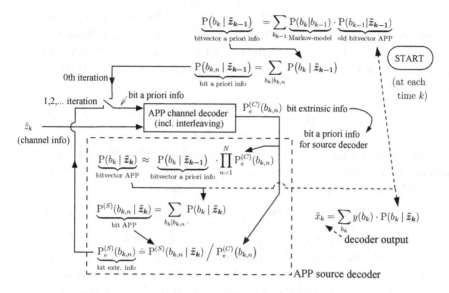

Fig. 3.6   Iterative source-channel decoding: algorithmic flowchart.

3.3.2   *Efficient implementation of ISCD by L-values**

The iterative source-channel decoder described in the previous section consists of two constituent decoders: an APP-algorithm for channel decoding and an APP-algorithm for source-decoding. An efficient implementation of ISCD by use of L-values [Hagenauer *et al.* (1996)] was stated in [Goertz (2001a)]; the scheme is depicted in Figure 3.7.

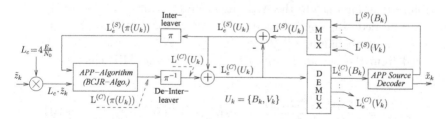

Fig. 3.7   Iterative source-channel decoding with APP decoders exchanging L-values (reprinted from [Hagenauer and Goertz (2003)] with permission, ©2003 IEEE).

*This section is based on "A Generalized Framework for Iterative Source-Channel Decoding" by N. Goertz, which appeared in *Annals of Telecommunications*, pp. 435–446, July/August 2001, ©2001 Annals of Telecommunications.

The soft-values that are passed between the APP decoders in Figure 3.7 are not the probabilities used in the previous section, but log-likelihood-ratios (L-values), which are directly related to them. For instance, the extrinsic L-value from the APP source decoder for the bit $b_{k,n}$ is defined[8] by

$$L_e^{(S)}(B_{k,n}) \doteq \log_e \frac{P_e^{(S)}(b_{k,n} = 0)}{P_e^{(S)}(b_{k,n} = 1)} \, , \qquad (3.38)$$

using the natural logarithm. The other L-values are related to the corresponding probabilities in a similar fashion. To simplify Figure 3.7, *vectors* $L(B_k)$ of L-values are used that are defined as follows:

$$L(B_k) \doteq \left\{ L(B_{k,n}), \ n = 1, ..., N_B \right\} \, . \qquad (3.39)$$

The advantage of L-values is that they cause less numerical problems than probability-values in the implementation of algorithms with finite word-length on a digital computer. The BCJR-algorithm, an efficient implementation of the APP-algorithm, e.g., for decoding of binary convolutional channel codes, can be completely carried out in the L-value domain (Log-MAP-algorithm, [Robertson *et al.* (1997)]). The use of such an algorithm is assumed in Figure 3.7. The received channel values $\tilde{z}_{k,l}$ are converted to L-values at the input of the APP channel decoder by multiplication with the factor $L_c = 4\frac{E_s}{N_0}$. This follows from the definition of the L-values, the PDF (3.3), and the assumption that the code-bits are uniformly distributed:

$$L(Z_{k,l}) = \log_e \frac{P(z_{k,l} = 0 | \tilde{z}_{k,l})}{P(z_{k,l} = 1 | \tilde{z}_{k,l})} = \log_e \frac{p_c(\tilde{z}_{k,l} | z_{k,l} = 0)}{p_c(\tilde{z}_{k,l} | z_{k,l} = 1)} = 4\frac{E_s}{N_0} \tilde{z}_{k,l} \, . \quad (3.40)$$

Since the APP source decoder, as it has been stated above, processes probabilities, an interface to the APP-algorithms operating in the L-value domain is needed: the computation of the bit-vector APPs by (3.35) requires the bit-probabilities $P_e^{(C)}(b_{k,n})$, $n = 1, ..., N_B$. The latter can be computed from the output L-values $L_e^{(C)}(B_{k,n})$ of the APP channel decoder by

$$P_e^{(C)}(b_{k,n}) = \frac{\exp\left(L_e^{(C)}(B_{k,n}) \right)}{1 + \exp\left(L_e^{(C)}(B_{k,n}) \right)} \cdot \exp\left(- L_e^{(C)}(B_{k,n}) \cdot b_{k,n} \right) \, , \quad (3.41)$$

which directly follows from the inversion of the L-value definition (3.38). Since in (3.35) the product over all these probabilities is computed for one

---

[8]We re-introduce the random variable B_k for the bit-vectors at the quantizer output, because later on in (3.41) we will have to distinguish between a particular choice $b_k \in \{0, 1\}$ and the L-value of the random variable B_k.

bit-vector, this operation can be simplified by inserting (3.41) into (3.35), i.e.,

$$P(b_k \mid \tilde{z}_k) = G_k \cdot \left[\prod_{n=1}^{N_B} \exp\left(- L_e^{(C)}(B_{k,n}) \cdot b_{k,n} \right) \right] \cdot P(b_k \mid \tilde{z}_{k-1}) , \quad (3.42)$$

with the normalizing constant

$$G_k \doteq \prod_{n=1}^{N_B} \frac{\exp\left(L_e^{(C)}(B_{k,n}) \right)}{1 + \exp\left(L_e^{(C)}(B_{k,n}) \right)} \qquad (3.43)$$

that does not depend on the particular choice b_k of the random bit-vector variable B_k. Now, the product in (3.42) can be turned into a summation in the L-value domain:

$$P(b_k \mid \tilde{z}_k) = G_k \cdot \exp\left(- \sum_{n=1}^{N_B} L_e^{(C)}(B_{k,n}) \cdot b_{k,n} \right) \cdot P(b_k \mid \tilde{z}_{k-1}) . \quad (3.44)$$

Thus, the L-values from the APP channel decoder can be integrated into APP source decoding without converting the individual L-values back to probabilities if (3.44) is used instead of (3.35). This is a simplification that also has strong numerical advantages. Additionally, the left-hand side of (3.44) is a probability that must sum up to one over all bit-vectors $b_k \in \mathcal{B}$. Hence, the constant G_k can be computed from this condition instead of (3.43).

The computation of new bit APPs (from (3.44)) within the iterations must be still carried out by (3.36), but the derivation of the extrinsic L-values $L_e^{(S)}(B_{k,n})$, that are generated by the APP source decoder, can be simplified, since (3.37) requires a division which is turned into a simple subtraction in the L-value-domain:

$$L_e^{(S)}(B_{k,n}) \doteq \log_e \frac{P^{(S)}(b_{k,n} = 0 \mid \tilde{z}_k)/P_e^{(C)}(b_{k,n} = 0)}{P^{(S)}(b_{k,n} = 1 \mid \tilde{z}_k)/P_e^{(C)}(b_{k,n} = 1)}$$

$$= L^{(S)}(B_{k,n}) - L_e^{(C)}(B_{k,n}) . \quad (3.45)$$

Thus, in the whole ISCD algorithm the L-values $L_e^{(C)}(B_{k,n})$ from the APP channel decoder are used and the probabilities $P_e^{(C)}(b_{k,n})$ are no longer required.

3.3.3 *Simulation Results for ISCD*

For the simulations we generated M parallel auto-correlated source signals by filtering Gaussian random samples by first-order recursive low-pass filters (coefficient $a = 0.9$, see Figure 3.2); we used 5-bit optimal (Lloyd-Max) scalar quantizers and the natural bit-mapping. The quantizer output bits were spread by a random-interleaver at each time k and afterwards they were commonly channel-encoded by a terminated rate-1/2 recursive systematic convolutional code (RSC-code, [Berrou and Glavieux (1996)]). The codewords were transmitted over the AWGN-channel.

In the first simulation we considered $M = 2$ signals and a terminated memory-2 rate-1/2 RSC-channel-code[9] (code rate $R_c = 10/24$). The results are depicted in Figure 3.8. The source SNR[10] (averaged over all M signals)

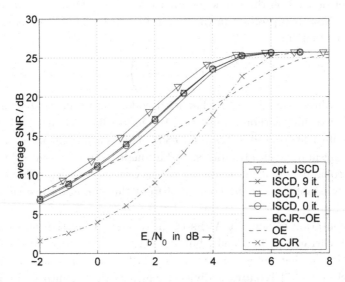

Fig. 3.8   Performance of iterative and near optimum joint source-channel decoding. Transmission of $M = 2$ strongly auto-correlated signals ($a = 0.9$) in parallel. Scalar quantization by 5 bits per index. Channel coding by a terminated memory-2 RSC code, rate $R_c = 10/24$. (Reprinted from [Goertz (2001a)] with permission, ©2001 Annals of Telecommunications)

is plotted vs. E_b/N_0, the ratio of the energy E_b per transmitted *data*-bit and the one-sided power spectral density N_0 of the channel-noise.[11]

---

[9]Generator polynomials: $g_0(D) = 1+D+D^2$ and $g_1(D) = 1+D^2$; $g_0(D)$ was used for the feedback part.

[10]For one of the M signals, the signal-to-noise ratio is computed by $\frac{SNR}{\text{dB}} = 10 \log_{10} \left(\frac{\sum_k x_k^2}{\sum_\kappa (x_\kappa - \tilde{x}_\kappa)^2} \right)$.

[11]The variance of the channel-noise samples in the discrete-time simulation equals $\sigma_n^2 = N_0/(2E_s)$ and $E_b/N_0 = \frac{1}{R_c} E_s/N_0$, where E_s is the energy that is used to transmit a

Clearly, the near optimum decoder denoted by "opt. JSCD" (see Section 3.2) works best, followed by the iterative decoding (curves "ISCD"). It should be noticed that the difference between the "opt. JSCD" and the "ISCD"-curves is only due to the approximations involved in the iterative decoder; the "opt. JSCD" is feasible in this case, just because $N \leq 10$.

We realized an additional decoder (curve "BCJR-OE") which used the BCJR-algorithm for channel decoding with "zero" a-priori information followed by optimal estimation (OE) (see end of Section 3.2) for source decoding, i.e., the scheme had the same constituent algorithms as ISCD. Due to its better initialization, ISCD outperforms the "BCJR-OE"-decoder, even if only the 0-th iteration[12] is carried out. In contrast to "BCJR-OE," ISCD utilizes the time-based correlations of the index-bits for channel decoding. Obviously, ISCD can not take advantage of the mutual dependencies of the index-bits, since the performance hardly increases with the number of iterations. This can be explained by the short length of the channel code which does not allow sufficient interleaving[13] to make the extrinsic bit-information from the source decoder "virtually" independent for the channel decoder.

It is known that the performance of concatenated codes with bit interleaving and their iterative decoding schemes is better, if a code with "long" codewords is used. Therefore, a similar simulation as above was carried out, but with $M = 50$ commonly channel-encoded indices and a terminated rate-1/2 memory-4 RSC-code[14] (code rate $R_c = 250/508$). The results are depicted in Figure 3.9: now, the first iteration leads to a strong improvement compared to "ISCD, 0 iter." and "BCJR-OE", i.e., the mutual correlations of the bits can be exploited in part by the iterative decoding scheme. Unfortunately, the near optimum decoder is too complex in this case (due to the block length), so a comparison with the best possible performance of the concatenated code (without the approximations inherent in ISCD) can not be carried out. It is interesting to notice that more than one iteration does not significantly increase the SNR in Figure 3.9. The reason is that the natural quantizer bit-mapping is not well suited for use in ISCD, as we will see in the next section.

---

channel-code bit.

[12]To avoid confusion with the counting of iterations, we start to count at zero, i.e., the 0-th iteration is one step of APP channel decoding followed by one step of APP source decoding.

[13]One should recall that the length of the interleaver is limited to the number of information bits at each time k in order to avoid additional delay.

[14]Generator polynomials: $g_0(D) = 1 + D^3 + D^4$ and $g_1(D) = 1 + D + D^3 + D^4$; $g_0(D)$ was used for the feedback part.

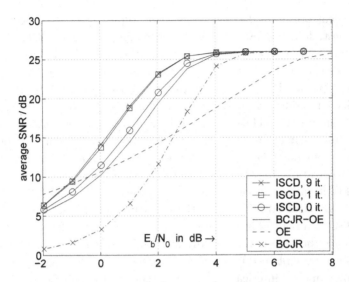

Fig. 3.9   Performance of iterative source-channel decoding and its constituent algorithms. Transmission of $M = 50$ strongly auto-correlated signals ($a = 0.9$) in parallel. Scalar quantization by 5 bits per index. Channel coding by a terminated memory-4 RSC code, rate $R_c = 250/508$. (Reprinted from [Goertz (2001a)] with permission, ©2001 Annals of Telecommunications)

In Figure 3.10, the results of a simulation similar to the previous one are shown. The difference is that the source signal is only weakly auto-correlated by setting the low-pass-filter coefficient to $a = 0.5$. This time, there is almost no difference between "BCJR-OE" and "ISCD, 0 iter.," which means that the *bits* are almost uncorrelated (in contrast to the indices). Qualitatively, the behavior of ISCD is the same as above. The first iteration produces a gain (a small one in this case) but two or more iterations again do not further increase the performance. Here, the gain by the first iteration is due to the well known [Jayant and Noll (1984)] non-uniform probability distribution of the indices at the output of an optimal quantizer, which leads to bit-correlations in the quantizer bit-vectors that can be exploited by the iterations.

In Figure 3.11, the influence of the interleaver-length on the performance of ISCD is investigated. For this purpose a terminated memory-6 rate-1/2 RSC-code[15] was used and the source signals were quantized by 4 bits. The

---

[15]Generator polynomials: $g_0(D) = 1+D^2+D^3+D^5+D^6$ and $g_1(D) = 1+D+D^2+D^3+D^6$; $g_0(D)$ was used for the feedback part.

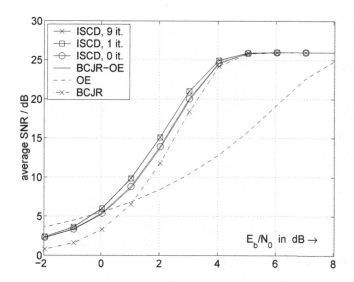

Fig. 3.10 Performance of iterative source-channel decoding and its constituent algorithms. Transmission of $M = 50$ weakly auto-correlated signals ($a = 0.5$) in parallel. Scalar quantization by 5 bits per index. Channel coding by a terminated memory-4 RSC code, rate $R_c = 250/508$. (Reprinted from [Goertz (2001a)] with permission, ©2001 Annals of Telecommunications)

filter-coefficient in Figure 3.2 was set to $a = 0.9$, i.e., the signals were strongly auto-correlated. The performance of ISCD was compared for $M = 15$ indices (resulting in 60 data bits) and $M = 100$ indices (resulting in 400 data bits), in both cases with a random interleaver and without an interleaver. Figure 3.11 indicates that the application of an interleaver significantly improves the performance of ISCD. The gain[16] gets larger with increasing block-length.

3.4 Quantizer Bit Mappings for ISCD

3.4.1 *Basic Considerations*

If we assume a low-pass correlation for the input, the source sample x_k will be located close to x_{k-1}. Thus, if the input at time $k-1$ is scalar quantized

---

[16]Even if no interleaver is used, a gain caused by the increasing block-length can be observed. The reason is that the rate-loss due to the tail-bits of the terminated convolutional code is smaller if the block-length is enlarged (E_b/N_0 is noted on the x-axis, where E_b is the energy per transmitted *data*-bit).

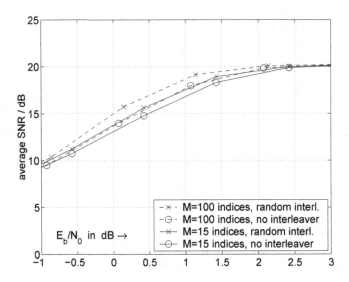

Fig. 3.11   Performance of iterative source-channel decoding (with 9 iterations) in dependence on the interleaver length. Transmission of $M = 15$ and $M = 100$ strongly auto-correlated signals ($a = 0.9$) in parallel. Scalar quantization by 4 bits per index. Channel coding by a rate-1/2 terminated memory-6 RSC code (code rate $R_c = 60/132$ for $M = 15$ and $R_c = 400/812$ for $M = 100$). (Reprinted from [Goertz (2001a)] with permission, ©2001 Annals of Telecommunications)

by, e.g., y_1 (see Figure 3.12), the next quantized value at time k will be y_0, y_1, or y_2 with high probability, while the probability for, say, y_7 is small. This is illustrated by Figure 3.13. The indices i_k of the quantizer levels y_{i_k} are mapped to bit-vectors $b_k = \gamma(i_k)$ as depicted in Figure 3.12.

If the channel code is strong enough so that hard decisions for the data bits could be taken correctly with high probability, we can idealize this situation by assuming that the (soft) a-priori information for the source decoder is perfect. Within the iterative decoding scheme this simply means that the APP source decoder tries to generate extrinsic information for a particular data bit, while it knows all other bits exactly. The situation is also illustrated by Figure 3.12: we assume that a three-bit scalar quantizer with the reproduction levels $y_0, ..., y_7$ is used for encoding of low-pass correlated source samples x_k. The quantizer levels are given three different bit mappings: natural binary, Gray, and optimized for ISCD. As an example, we consider the case that the bit-vector of the quantizer reproduction level y_1 has been transmitted and that the two leftmost bits are known (both

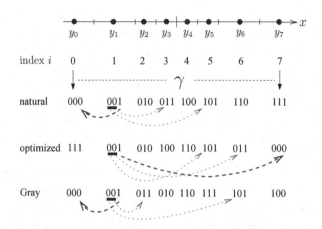

Fig. 3.12 Bit mappings for a three-bit quantizer ($N_B = 3$). (Based on [Hagenauer and Goertz (2003)] with permission, ©2003 IEEE)

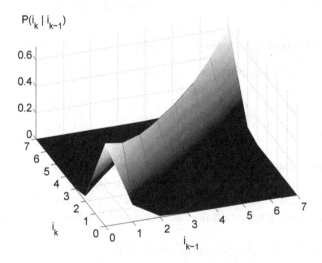

Fig. 3.13 Conditional probability distribution of successive three-bit Lloyd-Max quantizer indices for a strongly low-pass-correlated Gaussian source.

are zero in all mappings), due to the a-priori information from the channel decoder. We now try to generate extrinsic information for the rightmost bit from the source redundancies.

If we use the natural or the Gray mapping and flip the rightmost bit, we end up with quantizer level y_0 instead of y_1. Since y_0 and y_1 are neighbors in the source signal space we cannot decide with low error probability whether the rightmost bit is "one" or "zero," because both y_0 *and* y_1 are highly probable due to the low-pass correlation of the source samples.

The situation is different if we use the optimized mapping: since we jump to quantizer level y_7 (which is highly improbable) if we flip the rightmost bit, we can take a safe decision in favor of "one." Thus, the extrinsic information generated by the APP source decoder is strong and it will aid the channel decoder in the next iteration.

The example above suggests the concept to optimize the bit mapping: we have to allocate bit-vectors to the quantizer reproduction levels such that if we flip one of the bits, each pair of reproduction levels has a *large* distance in the source signal space. This still holds, if the source signal is high-pass correlated (with zero-mean and a symmetric probability density function): in this case, the source sample x_k will be close to $-x_{k-1}$ with high probability. Thus, the goal is again to maximize the distance between quantizer levels whose bit mappings differ in one bit position.

3.4.2 *Optimization by Binary Switching*

As argued above, the optimization criterion, the total "cost," is the expected distance

$$D \doteq \frac{1}{N_B} \sum_{e \in \mathcal{B}: w(e)=1} \mathrm{E}_{I|e}\Big\{ d(y_I, \underbrace{y_{\gamma^{-1}(\gamma(I)\oplus e)}}_{\stackrel{\mathrm{abb.}}{=} y(\gamma(I)\oplus e)}) \Big\} \qquad (3.46)$$

between any two quantizer levels whose bit-mappings differ in exactly *one* bit position, averaged over all N_B possible bit positions. Thus in (3.46), the Hamming-weight $w(e)$ of the "error" bit-vector e is fixed to *one*. The symbol "$\oplus$" denotes bitwise modulo-2 addition and γ is the bit mapping to be optimized.

In (3.46) we deliberately did not specify the distance measure $d(\cdot, \cdot)$, as we cannot give a clear preference for a particular one from a theoretical point of view. We will use the Euclidean distance and the square of it, the mean squared error, and compare both in the simulations.

The expectation in (3.46) for a particular "error" bit-vector e can be computed as follows:

$$\mathrm{E}_{I|e}\Big\{ d\big(y_I, y(\gamma(I) \oplus e)\big) \Big\} = \sum_{i=0}^{2^{N_B}-1} d\big(y_i, y(\gamma(i) \oplus e)\big) \cdot \mathrm{P}(i), \qquad (3.47)$$

where $P(i)$ denotes the known unconditional probabilities of the quantizer indices. If we insert (3.47) into (3.46) we obtain

$$D = \sum_{i=0}^{2^{N_B}-1} C(i) \qquad (3.48)$$

for the total cost, with the individual costs

$$C(i) \doteq \frac{P(i)}{N_B} \sum_{e \in \mathcal{B}:w(e)=1} d\left(y_i, y(\gamma(i) \oplus e)\right) \qquad (3.49)$$

of the quantizer code-vectors y_i.

In principle, it is possible to maximize the total cost D with respect to the bit mapping γ by the brute-force approach (full search over all possible mappings), but even if only $N_B = 5$ bits are used for quantization, the number of different mappings is extremely large ($2^{N_B}! \approx 2.6 \cdot 10^{35}$). Hence, we use a numerical approach with reasonable complexity that is based on the "binary switching algorithm" (BSA). It was proposed in [Zeger and Gersho (1990)] for a "Pseudo-Gray" reordering of the code-vectors in vector quantizer codebooks in the context of channel-optimized quantization. For the BSA the quantizer levels are assumed to be given (e.g., by codebook training) and an initial bit mapping is either created randomly or it is derived from the ordering of the quantizer levels in the codebook: the row numbers of the quantizer levels are converted to dual numbers which are interpreted as bit-vectors (natural mapping).

The flowchart of the BSA, adapted to our case, is depicted in Figure 3.14. The bit-vector corresponding to the code-vector[17] with the *lowest* individual cost (3.49) is tried to be switched with the bit-vector of another code-vector, the "switch partner." The latter is selected such, that the increase of the total cost due to the switch is as large as possible. If no switch partner can be found for the quantizer level with the lowest cost (that means all possible switches result in a lower total cost), the bit-vector of the quantizer level with the second-lowest cost will be tried to switch next. This process continues until a quantizer level from the list, sorted by increasing costs (3.49), is found that allows a switch of two bit-vectors that increases the total cost. After an accepted switch, the cost of each quantizer level and the total cost are recalculated, a new ordered list of quantizer levels is generated, and the algorithm continues as described above, until no further increase of the total cost is possible.

---

[17] The term "code-vector" is equivalently used for "quantizer level" to point out that the optimization scheme also works for vector quantizers.

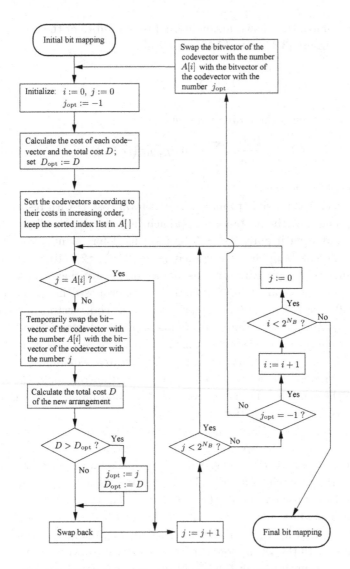

Fig. 3.14   Flowchart of the binary switching algorithm (BSA).

The only known upper bound for the number of switch-trials required to reach an optimum is trivial and extremely weak, as it is the total number $2^{N_B}!$ of different bit mappings [Zeger and Gersho (1990)]. In practice the BSA converges much faster: a few hundred up to thousand accepted switches are usually required for $N_B < 10$.

3.4.3 *Simulation Results with Optimized Bit Mappings*

As in Section 3.3.3, we generated auto-correlated source signals by process-ing Gaussian samples with the low-pass filter $H(z) = \frac{z}{z-a}$, with $a = 0.9$. Fifty of these auto-correlated but mutually independent signals were source encoded, each by a 5-bit optimal (Lloyd-Max) scalar quantizer, and the quantizer output bit-vectors were random-interleaved and jointly channel encoded by a memory-4 rate-1/2 recursive systematic convolutional code, which was terminated after each block of 50 bit-vectors (250 bits). Due to termination, the true channel code rate equals $R_c = \frac{250}{508}$, so over all the system had $5/R_c = 10.16$ channel-uses per source sample. The chan-nel codewords were transmitted over a binary-input AWGN channel and iterative source-channel decoding was performed at the decoder.

We optimized the quantizer bit mapping by binary switching in which we used the Euclidean distance as a distortion measure $d(\cdot, \cdot)$ in (3.49). The simulation results are depicted in Figure 3.15. In the realistic oper-ating range $(10\log_{10}(E_b/N_0) > 0\,\mathrm{dB})$, the performance of ISCD with the optimized bit mapping is much better than for the other mappings; com-pared to the Gray mapping and the natural mapping (the latter was used for the simulations in Section 3.3.3), the transmission power can be reduced by up to 1 dB at the same source SNR. The full gain is almost reached after four iterations; it takes, however, at least one iteration before the optimized mapping works better than the other ones. For very low E_b/N_0, the Gray mapping works best, because the channel decoder cannot generate reliable extrinsic information, so the assumption for our optimization criterion is not fulfilled that all bits of a bit-vector are known with high probability, excluding only the bit under consideration.

In Figure 3.16 we compare the performances of optimized bit mappings, when in (3.49) both the Euclidean distance (curve taken from Figure 3.15) and the mean squared error are used as the distance measures. Although the performances of both mappings after the first iteration are somewhat similar, we observe that until the first iteration the use of the mean-squared-error distance measure in the optimization leads to better results (especially in iteration "zero"), but eventually the use of the Euclidean distance mea-sure in the BSA results in a bit mapping with slightly better performance. As our goal is to achieve good performance for $10\log_{10}(E_b/N_0) > 0\,\mathrm{dB}$, we will use the bit mappings optimized with the Euclidean distance measure

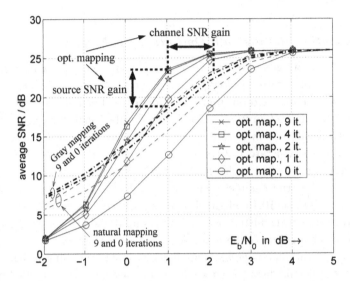

Fig. 3.15   Performance of ISCD for several bit mappings of scalar 5-bit quantizers for correlated Gaussian sources ($a = 0.9$). $M = 50$ indices are jointly channel encoded by a terminated memory-4 RSC code with rate $R_c = 250/508$. The optimized bit mapping was generated by binary switching with the Euclidean distance measure. (Based on [Hagenauer and Goertz (2003)] with permission, ©2003 IEEE)

in the following.

In Figure 3.17 we compare the performances of the natural bit mapping and an optimized bit mapping (Euclidean distance) for a 6-bit vector quantizer with dimension two, i.e., the quantizer bit rate equals 3 bits per sample. Again we jointly channel encoded $M = 50$ mutually independent but strongly auto-correlated vector-quantized source signals ($6 \cdot 50 = 300$ data bis in this case) by the terminated memory-4 rate-1/2 recursive systematic convolutional code. Hence, the true code rate is $R_c = \frac{300}{608}$, so over all we have $\frac{6}{2}/R_c = 6.08$ channel-uses per source sample. The simulation results are depicted in Figure 3.17. Although the gain due to the optimization of the bit mapping is somewhat smaller than for scalar quantization, it is still strong as we have a maximum source-SNR-gain of about 2 dB and a gain in "channel SNR" of about 0.8 dB.

It is interesting to compare the results for ISCD with the information theoretical limits. For scalar and vector quantization as described above, the best-performance curves of ISCD with the optimized bit mappings are

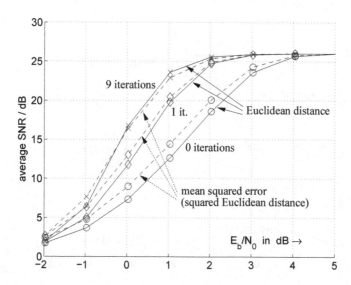

Fig. 3.16 Comparison of optimized bit mappings (scalar quantizer, 5 bits per index) for ISCD using the Euclidean distance and the mean-squared error in the optimization. $M = 50$ strongly correlated Gaussian source signals ($a = 0.9$) are jointly channel coded by a terminated memory-4 RSC code with a code rate of $R_c = 250/508$.

taken from Figures 3.15 and 3.17 and they are inserted[18] into Figure 3.18, along with the OPTA-curves (for details, see Appendix A, Section A.4.2). The latter describe the optimal performance theoretically attainable for the transmission of an auto-correlated Gaussian source ($a = 0.9$) which is transmitted over an AWGN channel with binary input.

We observe a rather large gap of about 4.5 dB in transmission power between ISCD and the performance limits. One reason is that the source-coding block-length is very small (for scalar quantization it is only one), because source coding is performed separately for each of the $M = 50$ mutually independent source signals. As mentioned in the introduction in

---

[18]In the Figures 3.15–3.17 we plot the performance vs. E_b/N_0, where E_b is the energy per transmitted *data* bit, which allows to compare systems with and without channel coding in a fair way (with respect to transmission power). In Figure 3.18, however, we plot the system performance vs. the ratio E_s/N_0, where E_s is the energy of a channel input symbol. This is reasonable, as we compare the system performance against the theoretical limits that do not implicate a fixed rate for channel coding. Hence, the curves from Figures 3.15, 3.17 have to be shifted by $10 \log_{10}(1/R_c) \approx 3$ dB towards negative dB-values of E_s/N_0.

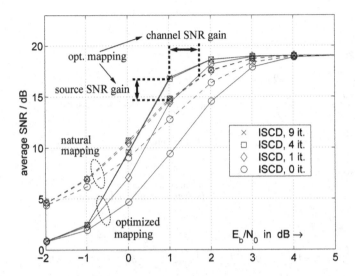

Fig. 3.17   Performance of ISCD with a two-dimensional 6-bit vector quantizer for each of $M = 50$ jointly channel-encoded mutually independent but strongly auto-correlated Gaussian sources ($a = 0.9$). The bit mapping was optimized for ISCD with the Euclidean distance measure. A terminated rate-1/2 memory-4 RSC code with rate $R_c = 300/608$ was used.

Section 3.1, the reason why we use this system model is that practical transmission schemes for multimedia signals apply source codecs, which extract parameters from the actual source signal that usually describe structurally different (and, hence, widely independent) properties of the signal such as spectral shape or energy. These parameters, which we interpret as our source signals, are often correlated in time. Although differential parameter encoding could be used to reduce the bit rate, this is often avoided because of the resulting error propagation problems in error-prone transmission environments.

Another reason for the large distance to the performance limits in Figure 3.18 is the relatively small but realistic block-length for channel coding. If the block-length could be enlarged[19] (i.e., more bit vectors were jointly encoded), we could expect some gain by applying turbo codes [Berrou and

---

[19]The block length for channel coding mainly depends on the source codec that is used. In narrow-band speech coding, for instance, the typical bit rate lies around 10 kbits/s. Due to the delay constraint of about 20 ms, we obtain a maximum block-length of 200 bits for channel coding.

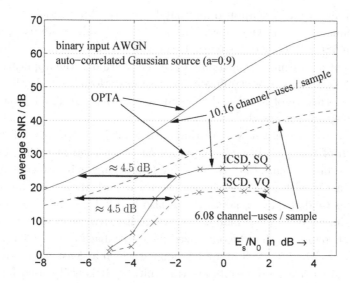

Fig. 3.18   Comparison of ISCD with scalar/vector quantization, rate-1/2 channel coding, and optimized bit mappings with the theoretical performance limits (OPTA).

Glavieux (1996); Hagenauer *et al.* (1996)]. The resulting coding system would concatenate three code components (i.e., two convolutional channel codes and the source redundancies), so the decoding process would be very complex, as three component decoders had to interchange extrinsic information. In [Zimmermann (2003)], this topic was investigated to some extent. For a strongly correlated source, the results for ISCD with a turbo code at a block-length of 1000 input bits show a gain of about 1 dB in transmission power over ISCD with a convolutional code; the complexity is, however, much higher, as more iterations (about ten) are required, and each iteration is made up of three APP-decoding steps.

3.5   Conclusions

We discussed iterative source-channel decoding (ISCD) for communication systems that use binary channel codes. ISCD is derived from a near optimum joint source-channel decoding algorithm by insertion of an approximation that allows to decompose the scheme into two constituent algorithms, namely the well-known symbol-by-symbol APP-algorithm for channel decoding and an newly defined APP source decoder. As in decoding of turbo codes, both algorithms are applied alternately in several iterations.

In the simulations, in which convolutional channel codes were used, we showed that ISCD achieves a strongly better quality of transmission than its constituent algorithms applied only once. Furthermore, the application of an interleaver leads to a significant increase in performance. Both the absolute performance of ISCD and the gain due to the interleaver grow if the block-length is increased.

Moreover, we stated a new algorithm with realistic complexity for the optimization of the quantizer bit mappings for application in ISCD. The simulation results indicate that again strong improvements in transmission quality are achieved for both scalar and vector quantization. As the optimization method is quite general, it can also be used in other scenarios such as bit-interleaved coded modulation where iterative processing is applied and bit mappings shall be optimized [Schreckenbach *et al.* (2003)].

Over all, ISCD works well in applications in which at least moderately long blocks of correlated data bits are transmitted. It is well suited for a combination with practical source-codecs for speech, audio, or image signals, because it exploits the residual redundancies in a systematic way and it allows to combine arbitrary scalar or vector quantization schemes with freely selectable bit rates in one joint decoding algorithm.

Moreover, ISCD offers scalable complexity at the decoder, since it is up to the system designer how many iterations shall be performed.

Chapter 4

Channel-Adaptive Scaled Vector Quantization*

4.1 Introduction

In Section 2.5.3, we briefly discussed channel-optimized vector quantization (COVQ) [Kumazawa *et al.* (1984); Zeger and Gersho (1988); Farvardin (1990); Farvardin and Vaishampayan (1991)] and we showed that it achieves strong quality-improvements over conventional vector quantization (VQ) if the transmission channel is noisy. Variations of COVQ in which simulated and deterministic annealing are used have been proposed, e.g., in [Gadkari and Rose (1999); Miller and Rose (1994)]; the algorithms work superior to "normal" COVQ due to improvements in the codebook design that avoid being trapped in poor local optima. Although a great deal of the work on COVQ has been done for the binary symmetric channel, the codebook training and the design of optimal encoder-decoder pairs for soft source decoding after the transmission with soft-decision demodulation have also been considered, e.g., in [Alajaji and Phamdo (1998); Skoglund (1999)]. Recently, the adaptation of COVQ to time-varying channels has become a subject of interest [Jafarkhani and Farvardin (2000)]; COVQ design approaches are also proposed for different types of incomplete channel information.

In this chapter, a new memory-efficient COVQ approximation for *time-varying* channels is presented. More precisely, we address the problem of how to limit the memory and complexity requirements in a COVQ framework such that these are almost the same as in the conventional VQ case, while keeping good performance for all (time-varying) channel conditions. It is shown that this goal can be achieved by a COVQ reference codebook

---

*Reprinted from *EURASIP Signal Processing*, vol. 83, N. Goertz and J. Kliewer, "Memory Efficient Adaptation of Vector Quantizers to Time-Varying Channels", pp. 1519–1528, ©2003, with permission from Elsevier.

that is scaled depending on the current bit error rate on the channel. A closely related method has been stated in [Ben-David and Malah (1994)], where scaling is introduced to decrease the distances between the code-vectors and thus to alleviate the quality decrease at the decoder output due to index permutations caused by bit errors. In our work, however, we employ code-vector-scaling to obtain a memory efficient approximation of a COVQ codebook that is designed for a bit-error probability being different from the one, the reference codebook is designed for.

After a discussion of complexity and memory issues for VQ and COVQ in Section 4.2, the new approximation of COVQ called channel-adaptive scaled vector quantization (CASVQ) is introduced in Section 4.3. In Section 4.4 the performance of the new scheme is compared with COVQ.

4.2   Memory and Complexity Issues for Vector Quantization (VQ) and Channel-Optimized VQ

Figure 4.1 shows the model of the transmission system that is used in this chapter. It consists of a channel-optimized vector quantizer with the codebook $\mathcal{Y}$, a binary symmetric transmission channel that causes index-permutations by bit errors, and a decoder that essentially performs a table-lookup. The bit-error probability is assumed to be known at both the encoder and the decoder.

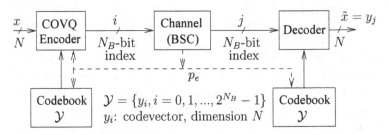

Fig. 4.1   Transmission system with a channel-optimized vector quantizer.

To simplify the notation below, we use the quantizer index i directly as the input for the channel, i.e., we treat the mappings from the quantizer indices to bit-vectors and vice versa as parts of the channel. The latter may be a composite that possibly contains channel coding, so in this case, p_e denotes the residual bit error rate after channel decoding. Thus, we use N_B-bit indices as channel inputs and not bit-vectors with K bits, as in our introductory discussion of COVQ in Section 2.5.3.

The idea of COVQ is to exploit the knowledge about the channel in the design of the quantizer codebook and in the encoding algorithm. The COVQ codebook design algorithm (see [Farvardin (1990)] and Section 2.5.3) is an extension of the classical design algorithm for noiseless channels (LBG-algorithm [Linde *et al.* (1980)]). In the following, we use the performance of COVQ as a reference for the approximations that are investigated.

The COVQ distance measure, which is used in both the codebook design procedure and the encoding operation, is different from that in conventional VQ. This is due to the fact that transitions from the transmitted indices i to some indices j at the receiver occur with probabilities $P_{j|i}$ that depend on the channel. As we transmit over a binary symmetric channel (BSC) with bit-error probability p_e, the transition probabilities are given by

$$P_{j|i} = p_e^{d_H(i,j)} \cdot (1 - p_e)^{N_B - d_H(i,j)} \quad , \qquad (4.1)$$

where N_B is the number of bits required to encode the index i and $d_H(i,j)$ is the Hamming distance between the bit-vectors of the transmitted and the received indices (implicitly assuming the natural binary bit mapping), i.e., $d_H(i,j)$ is the number of bit errors inserted by the channel. The probabilities $P_{j|i}$ are used in the COVQ encoder to minimize the expected distortion at the receiver output due to the quantization decision. The expected distortion, the *COVQ distance measure*, was already discussed in Section 2.5.3 (Equation (2.50)). It is given by

$$d_{\text{covq}}(x, y_i) = \sum_{j=0}^{N_Y - 1} d_{\text{vq}}(x, y_j) \cdot P_{j|i} \, , \qquad (4.2)$$

where x is the data-vector to be quantized and y_i the COVQ code-vector with the index $i = 0, 1, ..., N_Y - 1$. The number of code-vectors is denoted as $N_Y \doteq 2^{N_B}$ (size of the codebook), and $d_{\text{vq}}(x, y_j)$ is the distance measure that would be used in conventional VQ. As described in Section 2.5.3, the COVQ code-vectors y_i are the precomputed optimal estimates of the decoder output given by (2.40) which are also used for encoding by (4.2).

For COVQ encoding, (4.2) is computed for each code-vector y_i, and the index $i^\circledast$ corresponding to the code-vector that produces the minimum value $d_{\text{covq}}(x, y_{i^\circledast})$ is selected for the transmission over the channel. Besides, (4.2) is also used in the codebook training process. Often (and therefore also here) the mean-squared error

$$d_{\text{vq}}(x, y_i) = \frac{1}{N} \sum_{l=0}^{N-1} (x_l - y_{i,l})^2 \qquad (4.3)$$

is used as the *VQ distance measure*. The vector dimension is denoted by N and the vector components of y_i and x are indexed by l. The computation of (4.3) requires $3N$ floating-point operations per code-vector of dimension N, where "floating-point operation" will be used as a generic term for addition, subtraction, multiplication, or division in the following. It is a standard technique to simplify the implementation by expanding the sum in (4.3) as follows:

$$d_{vq}(x, y_i) = \frac{1}{N} \sum_{l=0}^{N-1} x_l^2 - \frac{2}{N} \sum_{l=0}^{N-1} x_l y_{i,l} + \frac{1}{N} \sum_{l=0}^{N-1} y_{i,l}^2 = \frac{1}{N} \sum_{l=0}^{N-1} x_l^2 + \frac{2}{N} d'_{vq}(x, y_i).$$
(4.4)

The first term is non-negative and does not depend on the code-vector. Since, furthermore, $\frac{2}{N}$ is a constant factor it is sufficient to minimize

$$d'_{vq}(x, y_i) \doteq q(y_i) - \sum_{l=0}^{N-1} x_l y_{i,l}$$
(4.5)

with

$$q(y_i) \doteq \frac{1}{2} \sum_{l=0}^{N-1} y_{i,l}^2$$
(4.6)

by a proper selection of the code-vector index i. In what follows, $d'_{vq}(x, y_i)$ will be referred to as the *simplified VQ distance measure*. The (halves of the) "energies" $q(y_i)$ of the code-vectors can be precomputed and stored in advance (since they do not depend on the input data x), so by use of N_Y additional scalar memory locations we can reduce the complexity of the codebook search to $2N$ floating-point operations per code-vector.

For COVQ, the calculation of the expected distortion (4.2) for each of the code-vectors requires the values of the VQ distance measure (4.3) for all code-vectors and the corresponding index transition probabilities $P_{j|i}$. Simplified implementations, which are algorithmically equivalent to the minimization of (4.2), have been stated in [Farvardin and Vaishampayan (1991)]: while the decoder uses the COVQ codebook directly, the encoder uses a "transformed" codebook that includes the index transition probabilities. We will briefly describe the method here: if we insert (4.3) into (4.2) and expand the sums we obtain

$$
d_{\text{covq}}(x, y_i) = \sum_{j=0}^{N_Y-1} \mathrm{P}_{j|i} \cdot d_{\text{vq}}(x, y_j) = \sum_{j=0}^{N_Y-1} \mathrm{P}_{j|i} \cdot \frac{1}{N} \sum_{l=0}^{N-1} (x_l - y_{j,l})^2
$$

$$
= \frac{1}{N} \sum_{l=0}^{N-1} x_l^2 + \frac{2}{N} \left\{ \underbrace{\sum_{j=0}^{N_Y-1} \mathrm{P}_{j|i} \frac{1}{2} \sum_{l=0}^{N-1} y_{j,l}^2}_{q(y_i)} - \sum_{l=0}^{N-1} x_l \underbrace{\sum_{j=0}^{N_Y-1} \mathrm{P}_{j|i} \cdot y_{j,l}}_{\overline{y_{i,l}}} \right\}.
$$

$$(4.7)$$

Thus, similar as in the VQ-case, it is sufficient to minimize the *simplified COVQ distance measure*

$$
d'_{\text{covq}}(x, y_i) = \overline{q(y_i)} - \sum_{l=0}^{N-1} x_l \, \overline{y_{i,l}} \,, \tag{4.8}
$$

with

$$
\overline{q(y_i)} \doteq \sum_{j=0}^{N_Y-1} \mathrm{P}_{j|i} \frac{1}{2} \sum_{l=0}^{N-1} y_{j,l}^2 \overset{(4.6)}{=} \sum_{j=0}^{N_Y-1} \mathrm{P}_{j|i} \, q(y_j) \tag{4.9}
$$

and

$$
\overline{y_{i,l}} \doteq \sum_{j=0}^{N_Y-1} \mathrm{P}_{j|i} \cdot y_{j,l} \,, \tag{4.10}
$$

instead of the minimization of (4.2). The "transformed" code-vectors $\overline{y_i} = \{\overline{y_{i,l}}, \, l = 0, ..., N-1\}$ from (4.10), which can be interpreted as the expectation of the code-vector at the receiver conditioned on the possibly transmitted index i, and the "energies" $\overline{q(y_i)}$ from (4.9) can be stored at the encoder in place of the actual codebook, which is only required at the decoder. Thus, the memory requirements and the complexity of COVQ encoding by minimization of (4.8) over i are not larger[1] than in the conventional VQ-case (where (4.5) is minimized)—as long as the channel is time-invariant.

If the channel is time-varying the channel-optimized codebook might not be matched to the current channel statistics (channel mismatch), i.e., the

---

[1] For COVQ on very noisy channels (e.g., bit error probabilities $p_e > 0.05$) empty encoding regions (corresponding to implicit error control coding) can be observed: certain indices are never selected by the encoder for transmission, i.e., the corresponding transformed code-vectors do not need to be considered in the distance computation (4.8). If many empty regions occur, this can lead to a significantly lower complexity [Farvardin and Vaishampayan (1991)] compared to the "standard" VQ-case.

performance degrades, compared to the optimal case. If the assumption of the bit error probability p_e for COVQ codebook training differs only slightly from its true value on the channel, the performance of unmatched COVQ is not significantly worse compared to optimally matched COVQ [Farvardin (1990)]. But if, for instance, a COVQ codebook designed for a bit-error probability of $p_e = 0.05$ is used on an uncorrupted channel ($p_e = 0$), a significant loss in the "clean-channel performance" can be observed.

One way to improve the performance in such a situation is to switch between a finite number of codebooks at the encoder and the decoder depending on the current channel state [Jafarkhani and Farvardin (1996)]; the codebooks can cover the range of possible channel variations (although no perfect match is possible since, in general, p_e is a real number). The robustness of COVQ against small variations of p_e preserves close-to-optimum performance if the number of codebooks is high enough. However, this strategy requires the storage of several codebooks at both the encoder and the decoder, which may be quite memory consuming. In the next section a new algorithm is stated that does not show this disadvantage by reducing the number of required codebooks to just one for all channel conditions.

4.3   Channel-Adaptive Scaled Vector Quantization

4.3.1   *Basic Principle*

In Figure 4.2, codebooks with $N_Y = 32$ two-dimensional code-vectors are depicted that were trained for a Gauss-Markov source (see Section A.2.3, Figure A.3) with the filter transfer function $H(z) = \frac{z}{z-a}$ and the coefficient $a = 0.9$. Such a source model is representative for the long-term statistics of many source signals as speech, audio, and images but also for the parameters that are extracted from a block of input samples by advanced source coding schemes. The expected mean-squared error (4.2) was used as a distance measure for COVQ codebook training. The plots show the code-vectors (marked by "×") that result from the training procedure for the COVQ codebooks [Farvardin and Vaishampayan (1991)] (with "splitting[2]" for the initialization) for several assumptions of the bit error probability p_e on a binary symmetric channel (BSC). For $p_e = 0$ the codebook is equal to a conventional VQ-codebook (Figure 4.2(a)).

---

[2]It is known from literature [Cheng and Kingsbury (1989)] that the splitting method in VQ-codebook training leads to a relatively good index assignment. In case of COVQ codebook training with $p_e > 0$ the index assignment is automatically optimized.

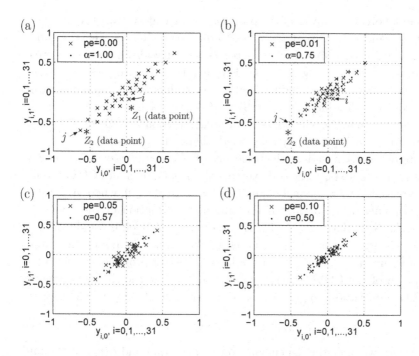

Fig. 4.2   Comparison of COVQ codebooks ("×") and CASVQ codebooks ("."), both with $N_Y = 32$ code-vectors. The codebooks in plot (a) are identical to a conventional VQ-codebook designed by the LBG algorithm [Linde *et al.* (1980)] (with "splitting").

Figure 4.2 shows that the code-vectors are placed closer to the "all-zero" vector (i.e., the mean of all training-data vectors) when the channel quality gets worse. This is because a permutation of a transmitted index i (which, for instance, may quantize the data point Z_1) to some other index j (caused by channel errors) leads to a large distortion, if the associated code-vectors have a large distance in the signal space; this is the case for the code-vectors marked by i and j in Figure 4.2(a). The distance between the vectors indexed by i and j in Figure 4.2(b) is smaller, i.e., the distortion caused by channel errors that permute both vectors is also smaller. On the other hand, the quantization of data-points with large distance from the origin (e.g., Z_2) is less accurate in Figure 4.2(b); thus, we have a loss in "clean-channel" performance ($p_e = 0$) using COVQ codebooks optimized for $p_e > 0$. In other words: a quality decrease due to higher quantizer overload-distortion is traded for the quality improvement due to a higher robustness against index permutations.

A more precise analysis of the positions of the code-vectors reveals (Figure 4.2(c) and 4.2(d)) that the code-vectors form clusters, i.e., the code-vectors are not only shrunk if p_e increases but they also have new relative locations. This clustering corresponds to empty coding regions that cause an implicit error control coding (redundant coding levels) as reported in [Farvardin and Vaishampayan (1991)]. However, a COVQ codebook for $p_e = p_1 > 0$ may be viewed as a shrunken version of COVQ codebook optimized for $p_e < p_1$, if only the "rough shape" is considered.

This observation leads to the basic idea of *channel-adaptive scaled vector quantization (CASVQ)*: the code-vectors $y_i^{(r)}$ from a reference COVQ codebook are scaled by a channel dependent factor $\alpha(p_e) \geq 0$, where $\alpha(p_e) \neq 1$ if the channel does *not* match the training assumption of the reference codebook. Note that $\alpha(p_e) < 1$ if the reference codebook is obtained for a smaller bit error probability compared to the actual one (as in the examples of Figure 4.2) and $\alpha(p_e) > 1$ if a larger bit error probability is used for the design of the COVQ reference codebook. Thus, the channel-matched COVQ code-vectors $y_i(p_e)$ are approximated by the CASVQ code-vectors

$$y_i'(p_e) = \alpha(p_e) \cdot y_i^{(r)} \, . \tag{4.11}$$

For brevity we omit the dependence on p_e for $y_i'(p_e)$ and $\alpha(p_e)$ in the following.

As an example, CASVQ code-vectors (marked by "·") are included in Figure 4.2. They have been derived from the VQ reference codebook (COVQ codebook with $p_e = 0$ as the design assumption) in Figure 4.2(a) by the scaling factors α stated in the legends. It is obvious that the rough shape of the COVQ codebooks (marked by "×") can be approximated by the CASVQ codebooks quite well.

4.3.2 *Optimization of CASVQ*

The question arises, how to find appropriate scaling factors for each channel condition, i.e., the function $\alpha = f(p_e)$ is required. Unfortunately, there is no way to find the optimal factor analytically by variational techniques, because the average distortion for a training-set of source vectors depends on the individual quantization decisions that again depend on the scaling factor. Hence, the quantization decisions change, as the scaling factor α is varied to minimize the average distortion. However, since the function $\alpha = f(p_e)$ is individual for a codebook and its source signal, it may be determined in advance before the system is used. Thus, we accept some computational complexity in the off-line optimization of the CASVQ code-book for a reasonable number of samples of α and p_e.

As an example, let us think of a correlated Gaussian source signal, which is vector-quantized by a codebook with $N_Y = 32$ two-dimensional code-vectors; (4.2) is used as a distance measure for the quantization and a conventional VQ codebook is used as a reference codebook for CASVQ. Thus, the range of the scaling factors is limited to $0 < \alpha \le 1$ in this case. The SNR-values, measured between the input of the quantizer and the output of a table-lookup decoder, are depicted in Figure 4.3 in a three-dimensional plot, for a simulation of several (α, p_e)-pairs. Figure 4.3 shows

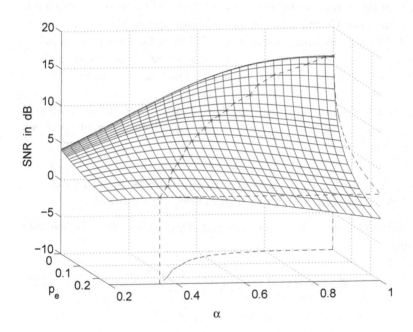

Fig. 4.3 Performance of CASVQ for codebooks with $N_Y = 32$ two-dimensional code-vectors which are used to encode a strongly correlated Gauss-Markov source ($a = 0.9$). The SNR-value is plotted versus the bit error probability p_e and the scaling factor α. A COVQ codebook for $p_e = 0$ (i.e., a conventional VQ-codebook) is used as a reference codebook.

that the SNR surface is quite smooth, i.e., a limited number of (α, p_e)-pairs is sufficient to achieve a good approximation of the best possible performance. The dashed curve on the SNR surface indicates the best SNR (and the associated α-value) for each simulated value of p_e; the required function $\alpha = f(p_e)$ is given by the projection of the dashed curve into the p_e-α-plane. The projection of the dashed curve into the SNR-p_e-plane indicates the best SNR-values achievable by CASVQ for each value of p_e. This curve appears again in the simulation results in Figure 4.5 (labeled "CASVQ, covq-dist.").

Note that the smooth behavior of the SNR surface in Figure 4.3 and of the function $\alpha = f(p_e)$, respectively, holds for arbitrary input sources. This is due to the fact that the COVQ code-vectors may be interpreted as pre-computed mean-square estimates of the vector-quantized transmitted source symbols. If the channel causes strong distortions the mean-square estimate at the output of the decoder tends to zero for any zero-mean input source, since the COVQ code-vector locations approach the zero vector. The CASVQ approach approximates this behavior for large p_e by $\alpha(p_e) \rightarrow 0$. Since the mean-square error between the source signal points and the CASVQ code-vector locations is a continuous function in α, the SNR surface and thus also $\alpha = f(p_e)$ are smooth functions.

By means of a simulation as described above, a close approximation of the function $\alpha = f(p_e)$ can be found for any practically relevant codebook in reasonable time.

4.3.3 *Complexity and Memory Requirements of CASVQ for Transmission over Time-Varying Channels*

If the CASVQ codebook is used, the simplified COVQ distance measure (4.8) gives

$$d'_{\text{covq}}(x, \alpha\, y_i^{(r)}) = \overline{q(\alpha y_i^{(r)})} - \sum_{l=0}^{N-1} x_l\, \overline{\alpha y_{i,l}^{(r)}}\,, \qquad (4.12)$$

with

$$\overline{q(\alpha y_i^{(r)})} = \sum_{j=0}^{N_Y-1} P_{j|i}\, \frac{1}{2} \sum_{l=0}^{N-1} \alpha^2\, (y_{j,l}^{(r)})^2 = \alpha^2 \cdot \sum_{j=0}^{N_Y-1} P_{j|i}\, q(y_j^{(r)}) = \alpha^2 \cdot \overline{q(y_i^{(r)})}$$
$$(4.13)$$

and

$$\overline{\alpha y_{i,l}^{(r)}} = \sum_{j=0}^{N_Y-1} P_{j|i} \cdot \alpha\, y_{j,l}^{(r)} = \alpha \cdot \sum_{j=0}^{N_Y-1} P_{j|i} \cdot y_{j,l}^{(r)} = \alpha \cdot \overline{y_{i,l}^{(r)}}\,, \qquad (4.14)$$

where (4.9) and (4.10) have been used. Note that both α and $P_{j|i}$ are functions of the *current* bit error probability p_e, which is distinct from the reference bit error probability. Thus, (4.12) may be written as

$$d'_{\text{covq}}(x, \alpha\, y_i^{(r)}) = \alpha^2 \cdot \overline{q(y_i^{(r)})} - \alpha \cdot \sum_{l=0}^{N-1} x_l\, \overline{y_{i,l}^{(r)}}\,. \tag{4.15}$$

As we can observe from (4.15) it is possible to adapt CASVQ to any value of p_e by use of an appropriate factor α and only *one* reference codebook for calculating the "transformed" reference code-vectors $\overline{y_i^{(r)}}$ and the "energies" $\overline{q(y_i^{(r)})}$ in (4.15). Whenever p_e changes, $\overline{y_i^{(r)}}$ and $\overline{q(y_i^{(r)})}$ have to be recomputed from the reference codebook by applying (4.9) and (4.10), respectively. Thus, the proposed CASVQ method with COVQ distance measure for encoding, which will be denoted with "CASVQ, covq-dist." in the following, is more complex than conventional VQ. Furthermore, memory for two codebooks, the reference codebook $y_i^{(r)}$ and the current transformed codebook $\overline{y_i^{(r)}}$ used for encoding, is required at the encoder. Note that the "energies" $\overline{q(y_i^{(r)})}$ only need to be stored for the current bit error probability, so here, no additional memory is required for time-variant channels. In contrast to that, a new codebook has to be trained and stored in both the encoder and the decoder for each value of p_e if we applied perfectly channel-matched COVQ.

In order to save the complexity for the re-computations (due to a change of p_e) and the memory for the second codebook in the CASVQ approach, we use the conventional VQ distance measure instead of the COVQ distance measure for encoding. Certainly, this is another approximation that will decrease the performance compared to the optimal COVQ; however, the degradation is only moderate but the simplification of the encoding is significant as we will see in the following. From (4.5) and (4.6) we obtain

$$d'_{\text{vq}}(x, \alpha\, y_i^{(r)}) = \underbrace{\frac{\alpha^2}{2} \sum_{l=0}^{N-1} (y_{i,l}^{(r)})^2}_{\alpha^2\,\cdot\, q(y_i^{(r)})} - \alpha \sum_{l=0}^{N-1} x_l y_{i,l}^{(r)} = \alpha^2 \left(q(y_i^{(r)}) - \sum_{l=0}^{N-1} \frac{x_l}{\alpha}\, y_{i,l}^{(r)} \right).$$

$$\tag{4.16}$$

Since α^2 in (4.16) is equal for each code-vector, it is sufficient to minimize

$$d''_{\text{vq}}(x, \alpha\, y_i^{(r)}) \doteq q(y_i^{(r)}) - \sum_{l=0}^{N-1} \left(\frac{x_l}{\alpha}\right) y_{i,l}^{(r)} \tag{4.17}$$

over i in place of (4.16). Now, all we have to do additionally (compared with conventional VQ) is to scale the components of the *input* data vector

x by $1/\alpha$ and to rescale the code-vector $y_j^{(r)}$ by α at the decoder output; a small table to represent the function $\alpha = f(p_e)$ is also required. The rest is the same as in conventional VQ including the memory and complexity requirements.

The CASVQ system with the VQ-distance measure for encoding is depicted in Figure 4.4. If this system is used, the corresponding simulation

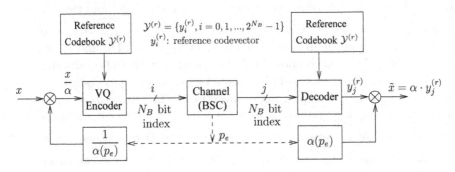

Fig. 4.4   CASVQ system with VQ distance measure for encoding.

results are labeled "CASVQ, vq-dist.". The simulation results in Section 4.4 show that the quality decrease due to replacing the COVQ distance measure with the simple VQ distance measure is only small.

Note that a change of α due to varying channel statistics affects both the encoder and decoder codebook. Therefore, we assume that a backward channel is available (which holds true for advanced wireless and wire-line communication systems) such that the estimated bit error rate at the decoder can be communicated to the encoder. If, for instance, a CELP speech codec [Schroeder and Atal (1985)] is used, its parameter quantizer codebooks could be adapted to the channel according to the CASVQ principle, and the scaled codebooks could be used in the closed-loop (analysis-by-synthesis) codec parameter quantization at the encoder.

4.4   Simulation Results

In Figure 4.5, the performances of COVQ and CASVQ for the transmission of a strongly auto-correlated Gauss-Markov source ($a=0.9$, see Figure 3.2) over a binary symmetric channel are depicted; the quantizations are carried out by codebooks with 32 two-dimensional code-vectors. The COVQ code-

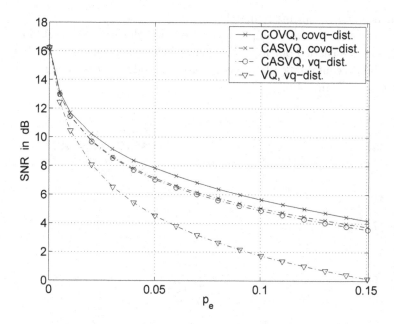

Fig. 4.5 Performance of COVQ, CASVQ, and VQ for the quantization of a strongly correlated Gauss-Markov source; codebooks with $N_Y = 32$ two-dimensional code-vectors are used.

books are optimally matched to the true value of p_e on the channel; the scaling factor $\alpha(p_e)$ of the CASVQ codebooks is adapted to p_e as described in Section 4.3.

As expected, the COVQ codebook with COVQ distance measure for encoding (curve labeled "COVQ, covq-dist.") works best of all, but there is only a moderate loss for "CASVQ, covq-dist.". Moreover, the performance of CASVQ with the conventional VQ distance measure for encoding ("CASVQ, vq-dist.") is only slightly inferior to "CASVQ, covq-dist.", i.e., the "better" distance measure does not significantly improve the performance of CASVQ.

The results for conventional VQ with VQ distance measure for encoding ("VQ, vq-dist.") have also been included in Figure 4.5 to evaluate the difference between COVQ and CASVQ: when a moderate loss in performance compared to optimally channel-matched COVQ can be accepted, it is possible to apply a CASVQ codebook with the simple VQ distance measure for encoding. This allows a very simple and memory-efficient adaptation of the coding scheme to time-varying channels, while keeping most of the performance-gain of optimally matched COVQ.

Qualitatively, the same implications hold for the second simulation, which again was carried out for the correlated Gaussian source signal, but with codebooks containing 128 three-dimensional code-vectors. The results are depicted in Figure 4.6.

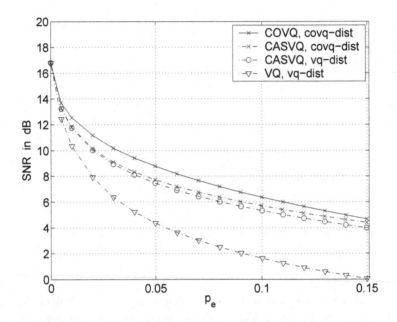

Fig. 4.6   Performance of COVQ, CASVQ, and VQ for the quantization of a strongly correlated Gauss-Markov source; codebooks with $N_Y = 128$ three-dimensional code-vectors are used.

The performances for an uncorrelated Gaussian source signal quantized with 32 two-dimensional code-vectors are shown in Figure 4.7. As above, CASVQ with the conventional VQ distance measure performs close to COVQ.

Figure 4.8 shows the dependencies of COVQ and CASVQ performance on the code-vector dimension N for a fixed number $N_Y = 64$ of code-vectors and a bit-error probability of $p_e = 0.05$. Since N is reciprocally proportional to the source coding rate it is clear that the SNR is essentially decreasing for increasing N. However, the largest gain of COVQ and CASVQ over conventional VQ is obtained for small vector dimensions. For strongly correlated ($a = 0.9$) source signals (where the use of vector quantization really pays off) the gain of CASVQ over VQ remains significant also for moderate vector dimensions. For uncorrelated source signals and high vector dimensions there is no significant gain by CASVQ over VQ, but in this case even COVQ works only moderately better than conventional VQ.

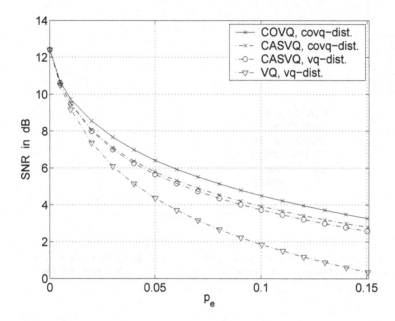

Fig. 4.7 Performance of COVQ, CASVQ, and VQ for the quantization of an uncorrelated Gaussian source; codebooks with $N_Y = 32$ two-dimensional code-vectors are used.

In Figure 4.9, in contrast to the previous simulation, the vector dimension is fixed ($N = 2$) but now the number N_B of quantizer index bits (and thus the codebook size $N_Y = 2^{N_B}$) is variable, again for a bit error probability of $p_e = 0.05$. For both strongly correlated ($a = 0.9$) and uncorrelated sources the gains of COVQ and CASVQ over VQ increase with the codebook size.

In summary, CASVQ works significantly better than conventional VQ, especially if the source signal is correlated and the bit rate is high or when the vector dimension is small and the number of code-vectors is large. The loss of CASVQ compared to COVQ gets larger with an increasing number of code-vectors, but at the same time the gain of CASVQ over conventional VQ increases.

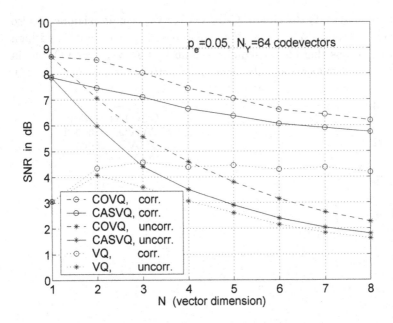

Fig. 4.8   Performance of COVQ, CASVQ (with VQ distance measure), and conventional VQ for the quantization of strongly correlated and uncorrelated Gaussian sources for various code-vector dimensions N; codebooks with $N_Y = 64$ code-vectors are used; the bit error probability is $p_e = 0.05$.

4.5   Conclusions

We have proposed channel-adaptive scaled vector quantization (CASVQ) as a substitute for channel-optimized vector quantization (COVQ). The advantage of CASVQ is that on time-varying channels the memory- and complexity requirements are practically the same as for conventional VQ, but, especially for correlated source signals, the performance is close to that of optimally channel-matched COVQ, which would require to store several codebooks for the adaptation to time-varying channels.

The CASVQ codebook is generated by scaling all the code-vectors in a reference codebook with a channel-dependent factor, which approximates the "shape" of the COVQ codebook for the current bit-error probability. If, additionally, the conventional VQ distance measure is used for encoding, the scaling can be moved from the code-vectors to the input source signal, yielding a further complexity reduction as then the normal VQ encoding algorithm can be used. The overhead induced by CASVQ with

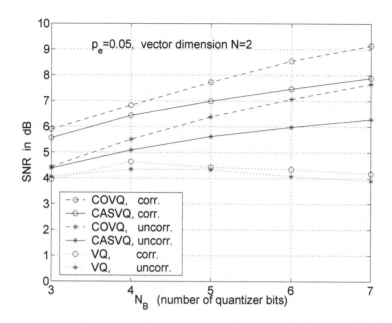

Fig. 4.9 Performance of COVQ, CASVQ (with VQ distance measure), and conventional VQ for the quantization of strongly correlated and uncorrelated Gaussian sources for various codebook sizes $N_Y = 2^{N_B}$; code-vectors with dimension $N = 2$ are used; the bit error probability is $p_e = 0.05$.

the VQ distance measure (compared with conventional VQ) is negligible, because CASVQ additionally requires just the scaling of the input source vector, rescaling after the transmission, and a small table to store appropriate scaling factors for the bit-error probabilities. Thus, CASVQ is very attractive for the transmission of multimedia signals (or the corresponding codec parameters) over time-variant channels, where limitations for complexity and memory exist, which do not allow the application of optimally channel-matched COVQ.

Chapter 5

Index Assignments for Multiple Description: Vector Quantizers*

5.1 Introduction

The principle of multiple descriptions (MD) is to represent a source signal by two or more descriptions, which are sent separately over erasure channels. The descriptions are generated in such a way that a basic quality is achieved at the decoder output from each individual description; the quality smoothly increases with the number of received descriptions.

The design of a multiple description quantizer can be divided into two parts: the selection of the quantizer reproduction levels (codebook training) and the choice of the index assignment. The latter is, in contrast to systems with a single description, a mapping from the indices of the quantizer reproduction levels to a *set* of descriptions.

The descriptions of an MD quantizer can be interpreted as the row- and column-indices of a matrix, in which the code-vectors, or equivalently their indices, are placed. The dimension of the matrix is denoted by K; it equals the number of descriptions that is selected by the system designer. The choice of the index assignments for multiple description quantizers may be seen as the problem, how to allocate the quantizer reproduction levels to the matrix cells in such a way that the distortion is minimized, when the descriptions are transmitted over an erasure channel. For multiple description *scalar* quantization (MDSQ) with two descriptions, optimal solutions for the index-assignment problem have been stated in [Vaishampayan (1993)]; the code-vectors are systematically placed along the main diagonal of the matrix. This concept makes use of the fact that a scalar quantizer codebook

---

*Based on "Optimization of the Index Assignments for Multiple Description Vector Quantizers" by N. Goertz and P. Leelapornchai, which appeared in *IEEE Transactions on Communications*, vol. 51, no. 3, pp. 336–340, March 2003. ©2003 IEEE.

may be ordered in such a way, that a reproduction level with a large value is allocated to a large index number in the quantizer codebook. Generally, such an ordering is impossible for a vector quantizer codebook.

Multiple description constraint *vector* quantizers with two descriptions have been studied, e.g., in [Vaishampayan *et al.* (2001); Goyal *et al.* (2002)], where a lattice structure is imposed on the quantizer codebook in order to allow for a systematic design of the index assignments. In [Fleming and Effros (1999)], an algorithm based on a ternary tree structure of the quantizer codebook is given for the design of multiple description vector quantizers with an arbitrary number of descriptions; the codebook and the index assignments are jointly optimized.

Compared with the algorithm in [Fleming and Effros (1999)], the work to be presented here is solely on the optimization of the index assignments of a multiple description vector quantizer; the quantizer codebook is assumed to be fixed and, in contrast to [Vaishampayan *et al.* (2001); Goyal *et al.* (2002)], without any structural constraints. This scenario is practically relevant, e.g., for packet based speech communication, where standardized speech codecs shall be used. In such a setup, the lossy part of the source encoder—e.g., the vector quantizer codebooks of the codec parameters—must not be changed, but the mapping of the encoder output bits to packets on the channel may be optimized.

The index assignment problem for a quantizer with a single description has been studied in [Zeger and Gersho (1990)]; there, the binary switching algorithm (BSA) is proposed, which minimizes the average system distortion, when the quantizer index is corrupted by bit-errors on the channel. The algorithm introduced here applies the idea of the BSA to optimize the index assignments of a multiple description *unconstrained* vector quantizer (MDVQ).

In what follows, we first state the model of an MDVQ system. Then, the optimal decoder is derived, and a useful formulation of the quality criterion is given for the optimization of the index assignments. Based on that, a new algorithm denoted by MD-BSA is introduced for the optimization of the index assignments of an MDVQ. Finally, the performance of the optimized MDVQ is discussed and compared with other multiple description quantizers and the rate-distortion bounds.

5.2 System Model

The block diagram of an MDVQ system with K descriptions is shown in Figure 5.1. The quantizer maps the input source vector[1] x to the nearest

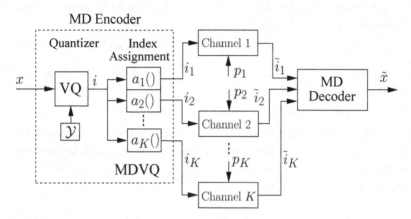

Fig. 5.1 Transmission system with a multiple description vector quantizer (MDVQ) with K descriptions.

code-vector y_i from the codebook $\mathcal{Y} = \{y_0, y_1, \ldots, y_{N_Y-1}\}$, where N_Y is the size of the codebook. The index assignments $i_l = a_l(i)$, $l = 1, \ldots, K$, map the quantizer output index

$$i \in \mathcal{I} \doteq \{0, 1, \ldots, N_Y - 1\} \tag{5.1}$$

to K descriptions i_l, $l = 1, \ldots, K$, that are transmitted over K memoryless and mutually independent channels, which cause packet erasures with the probabilities p_l, $l = 1, \ldots, K$.

Let M_l be the sizes of the index-sets of the descriptions $i_l, l = 1, \ldots, K$. Then the MD indices take values from the sets

$$i_l \in \mathcal{I}_l \doteq \{0, 1, \ldots, M_l - 1\}. \tag{5.2}$$

Since the index i must be uniquely de-codable from the *complete* set of descriptions, $N_Y \le M \doteq \prod_{l=1}^{K} M_l$ must be fulfilled. In most cases, $N_Y < M$ is selected, which is equivalent to adding redundancy, i.e., we have empty cells in the index assignment matrix (see Figure 5.2).

In what follows the optimal MDVQ decoder for a given codebook $\mathcal{Y}$ and given index assignments is stated. Then a new algorithm for the optimization of the index assignments $a_1(i), \ldots, a_K(i)$ is given.

[1] The source vectors x and the code-vectors y_i have dimension N; the components are assumed to be real numbers.

5.3   Optimal Decoder for a Given Index Assignment

The goal for the design of the decoder is to minimize the expected distortion at the output, for a set of received indices $\tilde{i}_1, ..., \tilde{i}_K$ (descriptions) and a given encoder with a fixed codebook and a fixed index assignment. Thus, the optimization problem can be stated as follows:

$$\tilde{x}(\tilde{i}_1, ..., \tilde{i}_K) = \arg \min_x \left\{ E_I \{ d(y_I, x) \,|\, \tilde{i}_1, ..., \tilde{i}_K \} \right\} . \tag{5.3}$$

In (5.3), the received indices are denoted by $\tilde{i}_l$; they take values from the sets

$$\tilde{i}_l \in \tilde{\mathcal{I}}_l \doteq \{ \varnothing, 0, 1, ..., M_l - 1 \}, \quad l = 1, ..., K , \tag{5.4}$$

which, in contrast to $\mathcal{I}_l$ according to (5.2), also contain the case of "erasure" indicated by "$\varnothing$". In what follows, the mean squared error will be used as a distortion measure $d(\cdot, \cdot)$. Therefore, the minimization (5.3) results in the minimum mean-square estimator

$$\tilde{x}(\tilde{i}_1, ..., \tilde{i}_K) = \sum_{i=0}^{N_Y - 1} y_i \cdot P(i \,|\, \tilde{i}_1, ..., \tilde{i}_K) , \tag{5.5}$$

(for comparison, see Section 2.3) where i denotes the quantizer index and y_i is the code-vector with the number i from the quantizer codebook $\mathcal{Y}$.

The main task of the receiver is to compute the a posteriori probabilities $P(i \,|\, \tilde{i}_1, ..., \tilde{i}_K)$. By use of the Bayes-rule we obtain

$$P(i \,|\, \tilde{i}_1, ..., \tilde{i}_K) = \frac{P(i, \tilde{i}_1, ..., \tilde{i}_K)}{P(\tilde{i}_1, ..., \tilde{i}_K)} = \frac{P(\tilde{i}_1, ..., \tilde{i}_K \,|\, i) \cdot P(i)}{P(\tilde{i}_1, ..., \tilde{i}_K)} , \tag{5.6}$$

where $P(i)$ denotes the known probability of the quantizer code-vector y_i. In the numerator we may equivalently insert the index assignments, i.e.

$$P(\tilde{i}_1, ..., \tilde{i}_K \,|\, i) = P(\tilde{i}_1, ..., \tilde{i}_K \,|\, i_1 = a_1(i), ..., i_K = a_K(i)) = \prod_{l=1}^{K} P(\tilde{i}_l \,|\, i_l = a_l(i)) , \tag{5.7}$$

where the rightmost equality is due to independence of the erasures on the channels. If we insert (5.7) into (5.6) we obtain:

$$P(i \,|\, \tilde{i}_1, ..., \tilde{i}_K) = \frac{1}{A} \cdot \prod_{l=1}^{K} P(\tilde{i}_l \,|\, i_l = a_l(i)) \cdot P(i) \tag{5.8}$$

with

$$P(\tilde{i}_l \,|\, i_l = a_l(i)) = \begin{cases} p_l & \text{if} \quad \tilde{i}_l = \varnothing \quad \text{(erasure)} \\ 1 - p_l & \text{if} \quad \tilde{i}_l = a_l(i) \\ 0 & \text{else} \end{cases} . \tag{5.9}$$

Equation (5.9) follows directly from the assumption for the channel that a packet is either erased with the probability p_l or it is received correctly. The constant A in (5.8) is defined by $A \doteq P(\tilde{i}_1, ..., \tilde{i}_K)$, but it is convenient to exploit that the left-hand side of (5.8) is a probability that must sum up to one over all possible $i = 0, 1, ..., N_Y - 1$. Therefore, A can be calculated by

$$A = \sum_{j=0}^{N_Y-1} \prod_{l=1}^{K} P\big(\tilde{i}_l \mid i_l = a_l(j)\big) \cdot P(j) \quad . \tag{5.10}$$

As an example, let us consider a system with $K=2$ descriptions. If the quantizer index i' is transmitted and the second description is erased by the channel, $\tilde{i}_1 = a_1(i')$ and $\tilde{i}_2 = \varnothing$ are received. Thus, we obtain from (5.8) and (5.9):

$$P\big(i \mid \tilde{i}_1 = a_1(i'), \tilde{i}_2 = \varnothing\big) = \frac{1}{A} \cdot P\big(\tilde{i}_1 = a_1(i') \mid i_1 = a_1(i)\big) \cdot p_2 \cdot P(i) \tag{5.11}$$

with

$$A = \sum_{j=0}^{N_Y-1} P\big(\tilde{i}_1 = a_1(i') \mid i_1 = a_1(j)\big) \cdot p_2 \cdot P(j) \overset{(5.9)}{=} (1-p_1) \cdot p_2 \sum_{j \in \mathcal{I}:\, a_1(j)=a_1(i')} P(j) \ . \tag{5.12}$$

Note that in (5.11) i' is the quantizer index that has been transmitted and that i is just a hypothesis for the transmitted index at the receiver side. In (5.12) a summation is carried out over all possible hypotheses for the quantizer index; these are denoted by j to avoid confusion in (5.13). If we combine (5.11) and (5.12) we obtain

$$P\big(i \mid \tilde{i}_1 = a_1(i'), \tilde{i}_2 = \varnothing\big) = \begin{cases} P(i) \Big/ \displaystyle\sum_{j \in \mathcal{I}:\, a_1(j)=a_1(i')} P(j) & \text{if} \quad a_1(i) = a_1(i') \\ 0 & \text{if} \quad a_1(i) \neq a_1(i') \end{cases}, \tag{5.13}$$

i.e., the so-called "side decoder" for the case that only the index i_1 has been received and i_2 is erased is given by (5.13) and (5.5). A similar formula results, if the first description is erased, i.e., $\tilde{i}_1 = \varnothing$ and $\tilde{i}_2 = a_2(i')$. Equation (5.13) indicates that all those indices i have to be used for the estimation of $\tilde{x}$ by (5.5) that have the same first description $a_1(i)$ as the one $\big(a_1(i')\big)$ that has been received. The denominator in the upper line of (5.13) normalizes the left-hand side, so that it becomes a probability. The decoding process is illustrated by Figure 5.2. When, for instance, description $\tilde{i}_1 = 0$ is received, but description $\tilde{i}_2$ is erased (Figure 5.2 a)), the "side"-decoder according to (5.13) takes the code-vector indices $i = 0, 2$ into account.

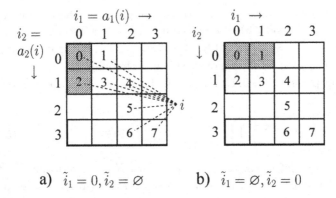

a) $\tilde{i}_1 = 0, \tilde{i}_2 = \varnothing$ **b)** $\tilde{i}_1 = \varnothing, \tilde{i}_2 = 0$

Fig. 5.2 Example with $K = 2$ descriptions that illustrates the decoding operation in the index assignment matrix, if $i = 0$ has been transmitted and a) description $\tilde{i}_2$ is erased, b) description $\tilde{i}_1$ is erased.

If both descriptions are received, the index a posteriori probability (5.8) equals

$$P\big(i \mid \tilde{i}_1 = a_1(i'), \tilde{i}_2 = a_2(i')\big) = \begin{cases} 1 & \text{if} \quad i = i' \\ 0 & \text{else} \end{cases} , \qquad (5.14)$$

so the "central" decoder issues $\tilde{x} = y_{i'}$ using (5.5). This is, for sure, the desired result for an uncorrupted transmission of the index i'. If no description is received, i.e., $\tilde{i}_1 = \varnothing$ and $\tilde{i}_2 = \varnothing$, we have $P\big(i \mid \tilde{i}_1 = \varnothing, \tilde{i}_2 = \varnothing\big) = P(i)$. Thus, the output of the receiver according to (5.5) equals the unconditional expectation of all code-vectors.

The optimal decoding operation described above is independent of the erasure probabilities, because the latter cancel out in (5.13). This still holds, if a system with more than $K = 2$ descriptions is used.

5.4 Quality Criterion for the Index Assignments

Similar as for the optimal decoder, the goal is to minimize the expected mean squared error, but, in contrast to (5.3), we now try to minimize it by the proper choices $a_l^{\circledast}()$, $l = 1, ..., K$, of the index assignments:

$$\{a_1^{\circledast}(), ..., a_K^{\circledast}()\} = \arg \min_{a_1(), ..., a_K()} \{D\} \qquad (5.15)$$

with

$$D \doteq \mathrm{E}_{I, \tilde{i}_1, ..., \tilde{i}_K} \big\{ d\big(y_I, \tilde{x}(\tilde{I}_1, ..., \tilde{I}_K)\big) \big\} . \qquad (5.16)$$

The optimal receiver output $\tilde{x}$ in (5.16) is computed as described in Section 5.3. The expectation (5.16) is unconditioned, since the index assignments shall be optimized only once, involving the statistics of the source *and* the channel. Equation (5.16) can be expanded as

$$D = \sum_{i=0}^{N_Y-1} \text{P}(i) \sum_{\tilde{i}_1 \in \tilde{\mathcal{I}}_1} \cdots \sum_{\tilde{i}_K \in \tilde{\mathcal{I}}_K} d\big(y_i, \tilde{x}(\tilde{i}_1, ..., \tilde{i}_K)\big) \cdot \text{P}\big(\tilde{i}_1, ..., \tilde{i}_K \big| i\big) ; \qquad (5.17)$$

the index sets $\tilde{\mathcal{I}}_l, l = 1, ..., K$, are given by (5.4). By insertion of the index assignments as in (5.7) and, due to the mutual independence of the erasures, (5.17) equals:

$$D = \sum_{i=0}^{N_Y-1} C(y_i) , \qquad (5.18)$$

with the "cost"

$$C(y_i) \doteq \text{P}(i) \sum_{\tilde{i}_1 \in \tilde{\mathcal{I}}_1} \cdots \sum_{\tilde{i}_K \in \tilde{\mathcal{I}}_K} d\big(y_i, \tilde{x}(\tilde{i}_1, ..., \tilde{i}_K)\big) \cdot \prod_{l=1}^{K} \text{P}\big(\tilde{i}_l \,\big|\, i_l = a_l(i)\big) \quad (5.19)$$

of each code-vector y_i. The probabilities in the product-term in (5.19) are given by (5.9).

As an example, we will again discuss the case of $K = 2$ descriptions. If (5.9) is inserted for the conditional probabilities into (5.19), the sums are expanded, and we consider that $\tilde{x}(a_1(i), a_2(i)) = y_i$ (i.e., $d(y_i, \tilde{x}(a_1(i), a_2(i))) = 0$), the costs of the code-vectors can be reformulated according to

$$C(y_i) = \text{P}(i) \bigg(d\big(y_i, \tilde{x}(a_1(i), \varnothing)\big) \cdot (1 - p_1) \cdot p_2 +$$

$$d\big(y_i, \tilde{x}(\varnothing, a_2(i))\big) \cdot p_1 \cdot (1 - p_2) +$$

$$d\big(y_i, \tilde{x}(\varnothing, \varnothing)\big) \cdot p_1 \cdot p_2 \bigg) . \qquad (5.20)$$

Now, the sum (5.18) may be split into two parts; the first one involves the first two terms from (5.20) that depend on the index assignments $a_1(), a_2()$, the last term is a positive constant for the optimization. If, additionally, the erasure probabilities on both channels are assumed to be equal, i.e., $p_1 = p = p_2$, (5.18) can be rewritten as follows:

$$D = \underbrace{(1 - p) \cdot p}_{\text{pos. const.}} \underbrace{\sum_{i=0}^{N_Y-1} \Delta C(y_i)}_{\doteq \Delta D} + \underbrace{p^2 \sum_{i=0}^{N_Y-1} \text{P}(i) \cdot d\big(y_i, \tilde{x}(\varnothing, \varnothing)\big)}_{\text{positive constant}}$$

$$(5.21)$$

with

$$\Delta C(y_i) \;=\; \mathrm{P}(i)\Big(d\big(y_i, \tilde{x}(a_1(i), \varnothing)\big) \;+\; d\big(y_i, \tilde{x}(\varnothing, a_2(i))\big)\Big). \qquad (5.22)$$

Thus, for the optimization of the index assignments it suffices to minimize ΔD. It is important, that ΔD is independent of the erasure probability p, i.e., for the case $K = 2$ and $p_1 = p = p_2$, the index assignments are optimal for all possible values of the erasure probability. That is *not* true if more than two descriptions $(K > 2)$ are used.

5.5   Optimization of the Index Assignments

5.5.1   *The Complexity Problem*

As illustrated by Figure 5.2, the selection of the index assignments can be seen as the problem, how to place N_Y indices into a K-dimensional matrix with $M \doteq \prod_{l=1}^{K} M_l > N_Y$ locations in such a way, that the distortion D given by (5.18) is minimized. The conceptually easiest way to do the optimization is the brute-force approach: one would "simply" have to compute the distortion for each possible index assignment and select the one with the lowest distortion. Since N_Y locations are taken out of M possible ones in the matrix and $N_Y!$ possible allocations of the code-vector indices exist for each choice of matrix locations, there are

$$\binom{M}{N_Y} \cdot N_Y! = \frac{M!}{(M - N_Y)!} \qquad (5.23)$$

possible assignments, i.e., the brute force approach is infeasible in all practically relevant cases: for example, if only $N_Y = 32$ code-vectors are mapped to two 3-bit descriptions, i.e., $M_1 = 2^3 = M_2$ and, thus, $M = 64$, the distortion of $4.8 \cdot 10^{53}$ different index assignments would have to be computed.

5.5.2   *Index Optimization by the Binary Switching Algorithm for a System with a Single Description*

The problem of assigning N_Y indices to N_Y code-vectors to control the performance degradation caused by bit errors on the channel is studied in [Zeger and Gersho (1990)], where the binary switching algorithm (BSA) is proposed to overcome the complexity problems of the brute-force approach, which would require to check $N_Y!$ different assignments.[2]

---

[2] As indicated by (5.23), the complexity problem is even worse in the multiple description case that is discussed here.

The basic idea of the BSA is to pick the code-vector with the highest cost (which has the strongest contribution to the total distortion) and try to switch the index of this code-vector with the index of another code-vector, the "switch partner." The latter is selected such that the decrease of the total distortion due to the index switch is as large as possible. If no switch partner can be found for the code-vector with the highest cost (that means all possible switches result in a higher total distortion), the code-vector with the second-highest cost will be tried to switch next. This process continues until a code-vector from the list, sorted by decreasing costs, is found that allows a switch that lowers the total distortion. After an accepted switch, the cost of each code-vector and the total distortion are recalculated, a new ordered list of code-vectors is generated, and the algorithm continues as described above, until no further reduction of the total distortion is possible. Along with a discussion, the flowchart of the BSA has already been stated in Section 3.4.2 in Figure 3.14; hence we will not go again into the details here.[3]

5.5.3 *Binary Switching for Multiple Descriptions*

In Section 5.4, the total distortion for the MDVQ system was formulated as the sum of the costs of the code-vectors. Hence, it is easy to adopt the idea of the normal BSA for multiple descriptions: as for the single-description case, an initial index assignment is used as a starting point for the multiple description BSA (MD-BSA), but the cost of each code-vector and the total distortion are calculated by (5.19) and (5.18), respectively.[4] The code-vectors are sorted according to their costs in decreasing order, and the candidate code-vectors for switching are picked from the top of the list. In contrast to the conventional BSA for single descriptions, now the switch partner can be either another code-vector or an empty location in the matrix that has not been assigned to any code-vector yet; this is illustrated by Figure 5.3.

The switch that is accepted achieves the lowest total distortion for the current candidate-code-vector from the list. After an accepted index switch, the cost of each code-vector and the total distortion are recomputed and a new list of code-vectors, ordered by decreasing costs, is generated for the

---

[3]The only differences are that we have to replace "$D > D_{opt}$" by "$D < D_{opt}$" in the "if"-statement and that the costs have to be sorted in decreasing order in the array $A[\,]$.

[4]For a system with two descriptions and equal packet erasure probabilities on both channels, the cost functions and the distortion can also be computed by (5.22) and (5.21), respectively.

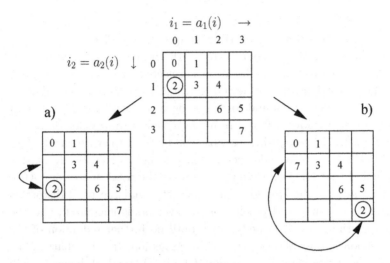

Fig. 5.3   Two possible ways of switching code-vectors in the index assignment matrix: a) switch with an empty position, b) switch with another code-vector.

next step. The algorithm stops, when no more switches are possible that further reduce the total distortion.

5.6   Simulation Results

A memoryless zero-mean unit-variance Gaussian source signal was used for the simulations. The VQ codebooks of size $N_Y = 64$ and 128 (quantizer indices of $N_B = 6$ and 7 bits) with a vector dimension of two were designed by the LBG algorithm; the splitting method [Linde *et al.* (1980)] was used for the initialization.

The quantizer indices were mapped to $K = 2$ descriptions,[5] each with 4

---

[5]Although the presented algorithm is able to optimize the index assignments in more general cases, $K = 2$ descriptions were used in the simulations because the rate-distortion bounds are *explicitly* known and well understood for this case and, as stated above, the result of the index optimization is independent of the particular value of the packet-erasure probability if both descriptions are independently erased with the same probability. Moreover, comparable numerical results for other MD vector-quantization schemes with more than two descriptions are not available in literature. To the best of our knowledge, [Fleming and Effros (1999)] is the only paper where SNR-plots (SNR vs. packet erasure probability) can be found for multiple descriptions with $K > 2$, but the results are given for magnetic resonance brain scan images, i.e., the source signal in [Fleming and Effros (1999)] is different from the Gaussian source model used in this work.

bits. For this purpose, a 16×16-matrix had to be filled with the indices of the code-vectors. For the initialization of the index assignment matrix for the MD-BSA and for the reference simulations we used two schemes:

(1) 1000 different random index assignments were tried and selected was the one with the lowest total distortion ("random initialization").
(2) the modified linear (ML) index assignment from [Vaishampayan (1993)] was used, which places the quantizer indices on the diagonals of the assignment matrix; therefore, this method is denoted by "diagonal initialization."

The ML-assignments were derived in [Vaishampayan (1993)] for multiple description *scalar* quantizers (MDSQ), where the indices have a direct relationship to the amplitudes of the quantizer reproduction levels. Since the splitting method was used for the initialization of the LBG codebook training, most of the neighboring indices of the VQ codebook lie also relatively close to each other in the signal space [Farvardin (1990)]. Therefore, the ML-assignment is useful for the MD-BSA as an initialization because it is already closer to a "good" assignment than some random initialization.

The descriptions were transmitted over mutually independent erasure channels with the erasure probabilities $p_1, p_2 \in (0...1)$. For both erasure probabilities the same values were always selected, i.e., $p_1 = p = p_2$, so the index optimizations were independent of the particular value of p, as stated for $K = 2$ in (5.21).

The performances of the initial and the optimized index assignments were compared by the SNR-values at the decoder output. The results in Figure 5.4 show that the MD-BSA achieves strong gains for both initializations. As expected, the ML-initialization works better than the random initialization, but the results after the optimizations are only slightly different.

It is interesting to compare the performance of the optimized MDVQ scheme with the rate-distortion bounds for multiple descriptions [Ozarow (1980)] and the MDSQ [Vaishampayan (1993)] for the Gaussian source with zero mean and unit variance. In Table 5.1, the side distortions of several schemes that all have the same central distortions[6] are compared.

---

[6]The performance of the vector quantizer without any erasures is described by the central distortion; it is a property of the quantizer that does not depend on the index assignment. The side distortions result at the decoder output if one of the channels always erases a description. If the side distortions are not the same for each description, they are called "unbalanced." In this case, one can use the average of the side distortions to obtain a single value as a figure of merit.

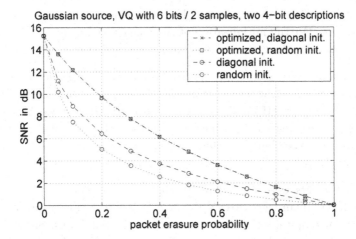

Fig. 5.4   Performance of MDVQ with the initial and the MD-BSA-optimized index assignments for a Gaussian source signal. The rate is 2 bits per source sample per description.

Table 5.1   Side distortions of the optimized MDVQ, the rate-distortion bound (OPTA) for multiple descriptions [Ozarow (1980)] and the MDSQ [Vaishampayan (1993)] for the same central distortions. All systems have two descriptions, each with a rate of 2 bits per source sample.

central distortion {central SNR/dB}	side distortion {side SNR/dB}		
	OPTA	MDVQ (dim. 2)	MDSQ
0.030 {15.2}	0.063 {12.0}	0.153 {8.2} (2^6 code-vec.)	0.190 {7.2}
0.015 {18.2}	0.083 {10.8}	0.256 {5.9} (2^7 code-vec.)	0.482 {3.2}

As stated above, the vector quantizers had a dimension of two and indices with 6 bits (central distortion of 0.03) and 7 bits (central distortion of 0.015). The indices were mapped to two 4-bit descriptions. The side distortions for the optimized MDVQ were measured and inserted into the table. At the same central distortions and for the same rates (2 bits per source sample per description) the OPTA-values[7] for the side distortions

---
[7]optimal performance theoretically attainable

were additionally picked from the rate-distortion region derived in [Ozarow (1980)]. The same was done for the multiple description scalar quantization (MDSQ) [Vaishampayan (1993)]: again for the same rates and central distortions, the values of the side distortions were picked from [Vaishampayan (1993)] (Figure 12) and they were inserted into the rightmost column of Table 5.1. Within the brackets, the SNR-values corresponding to the distortions were added in the whole table.

Table 5.1 indicates that the MDVQ with MD-BSA index optimization achieves significant gains (1 dB in side SNR for the higher central distortion, 2.7 dB for the lower central distortion) over MDSQ. The gain is larger, if the redundancy (number of unused matrix locations) is small. In the example with the two-dimensional vector quantizer, the side distortion of the optimized MDVQ is, however, still more than 4–5 dB away from the rate-distortion bound (OPTA), which is due to the small block-length of the quantizer.

In a simulation of the transmission system, the gains in side distortion (indicated by Table 5.1) of the index-optimized MDVQ turn into maximum SNR improvements over MDSQ of 0.5 dB and 1.9 dB for the higher and the lower central distortions, respectively, both for an erasure probability of about $p = 0.2$.

5.7 Conclusions

The multiple-description binary switching algorithm (MD-BSA) was introduced for the optimization of the index assignments of multiple description vector quantizers (MDVQ) with an arbitrary number of descriptions.

The index assignments for MDVQ resulting from the MD-BSA significantly improve the signal-to-noise ratio at the receiving end compared with multiple description scalar quantizers; this was shown by simulations for a system with two descriptions and a Gaussian source signal.

In the MDVQ system model, the code-vectors of the quantizer are assumed to be fixed, i.e., the lossy part of the source coding scheme is not affected by the optimization. This allows to apply the MD-BSA to standardized speech, audio, and image codecs that shall be used for signal transmission over packet erasure channels.

Chapter 6

Source-Adaptive Modulation*

6.1 Introduction

In this chapter we deal with a communication system, in which continuous-valued source samples are quantized and transmitted by a digital modulation scheme that is adapted to the source samples. The adaptation of the modulation signal constellation to the quantizer codebooks and vice versa can be seen as an approach to joint source-channel *encoding*. In some of these algorithms [Vaishampayan and Farvardin (1992); Han and Kim (2001); Liu *et al.* (1993)] the codebooks and the modulation signal sets are designed commonly for a source and a channel, both with fixed statistics. The adaptation is done in an "average sense," i.e., after the design procedure the codebooks and the modulation sets are fixed.

In contrast to such a "conventional" system optimization, we state a new scheme called source-adaptive modulation (SAM) that selects each transmitted signal point dependent on the *current unquantized* source signal. The transmitted signal point is *not* restricted to be one out of the fixed set that is used by the conventional decoder for detection. Compared to traditional systems, the new algorithm requires only modifications at the transmitter, i.e., the idea is especially attractive for systems with a relatively small number of transmitters and a large number of receivers that are already in use and shall not be changed. Two examples are radio broadcast and the down-link in a mobile radio network.

After the introduction of the system model in Section 6.2, the basic idea of source-adaptive modulation is described for the example of M-PSK

---

*Parts of this chapter are based on "Source-Adaptive Shifting of Modulation Signal Points for Improved Transmission of Waveform-Signals over Noisy Channels" by N. Goertz, which appeared in *Proceedings International Conference on Acoustics, Speech, and Signal Processing*, vol. 3, pp. 2517–2520, May 2002. ©2002 IEEE.

modulation in Section 6.3. The mathematical derivation of the optimal location of the transmitted modulation signal point is stated in Section 6.4, along with a discussion of methods for a low-complexity implementation and some simulation results. In Section 6.5 we describe, how the SAM principle can be applied to general QAM modulation schemes and we compare the performance with that of a conventional system that does not use SAM.

6.2   Conventional System Model

Our goal is to transmit the continuous-valued samples x of a source signal to a destination over a noisy channel. For this purpose, we use the transmission system depicted in Figure 6.1. We have omitted the time index of

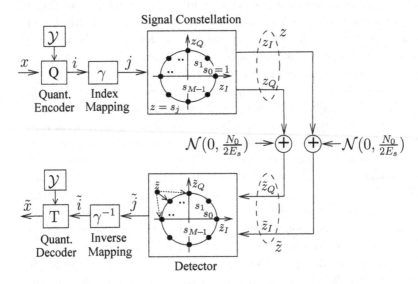

Fig. 6.1   System model (reprinted from [Goertz (2002)] with permission, ©2002 IEEE).

all signals, because the algorithms stated below operate independently for each realization of the input x. If correlations exist between the adjacent source signal samples they can be exploited by estimation-based decoders (see Chapter 3) that can be combined with the algorithms presented here.

In our system, the source signal x is first quantized.[1] In some illustra-

---

[1] A quantizer optimized for the probability density function of the source signal can be

tions that follow, x is assumed to be scalar to keep the discussion simple, but the use of vector source signals is, analytically and practically, straightforward, since only the values of the distance measure between the input signal and the decoded quantizer reproduction levels appear in the formulas below. As usual, the mean-squared error is used as a distance measure for the quantization, i.e.,

$$d(x, y_i) \doteq \frac{1}{N} \sum_{l=0}^{N-1} (x_l - y_{i,l})^2 \quad . \tag{6.1}$$

In (6.1), N is the vector dimension of the quantizer (the vector components are indexed by l) and y_i is a code-vector (quantizer reproduction level) from the codebook

$$\mathcal{Y} \doteq \{y_0, y_1, ..., y_{N_Y-1}\} \; ; \tag{6.2}$$

N_Y is the number of code-vectors, which for simplicity[2] is assumed to equal the number M of modulation signal points, i.e., $M = N_Y$. The quantizer output index i is mapped to another index

$$j = \gamma(i), \quad i, j \in \{0, 1, ..., M-1\} \;, \tag{6.3}$$

which is used to address the signal point s_j from the modulation signal constellation

$$\mathcal{S}_M \doteq \{s_0, s_1, ..., s_{M-1}\} \tag{6.4}$$

for transmission, i.e., $z = s_j$. As we will see later on, the mapping (6.3) has significant influence on the performance. If the bit error rate that results from the symbol permutations due to channel noise shall be minimized, a Gray-mapping is frequently used. As for us the bit error rate is not the quality criterion, other mappings will turn out to be more appropriate.

The M-PSK signal constellations are used as examples in Figure 6.1 and in some of the derivations below; the idea of SAM is, however, not restricted to that.

---

used, i.e., a Lloyd-Max quantizer [Jayant and Noll (1984)] for scalar inputs or a codebook trained by the generalized Lloyd algorithm [Linde *et al.* (1980)] for vector source signals. However, the SAM principle works with any other quantization scheme as well.

[2]If the quantizer has more reproduction levels than the modulation constellation has signal points, the most significant bits of the quantizer should be packed together onto one modulation signal point. Simulations have shown [Schimmel (2003)] that this scheme preserves most of the performance gains that we will report below. Due to the rather involved notation that is caused by splitting a quantizer index into bits and its transmission by several signal points, we will omit the details here.

We will assume that the transmission is carried out over an AWGN channel.[3] Since modulation schemes like M-PSK or QAM shall be used, the baseband signal z is split into an inphase component z_I and a quadrature component z_Q. Both can be treated as if they were transmitted over independent AWGN channels with the same noise variance $\sigma_n^2 = \frac{N_0}{2E_s}$ [Haykin (2000)], where E_s is the (average) energy of the transmitted channel symbols (energy sum of both channels) and $N_0/2$ is the noise power spectral density for both AWGN channel components. Note that, for the model and the simulations, the symbol power $P = E_s/T_s$ has been merged with the actual noise variance on the channel (for details, see Figure A.12 in Appendix A.3.3.4). If a simulation based on this model shall give a fair comparison in terms of E_s/N_0, it must be ensured that the average power P_n of the signal constellation is normalized to one, i.e.,

$$P_n \doteq \mathrm{E}\{\|Z\|^2\} = \mathrm{E}\{Z_I^2\} + \mathrm{E}\{Z_Q^2\} = \sum_{i=0}^{M-1} |s_{\gamma(i)}|^2 \cdot \mathrm{P}(i) \overset{!}{=} 1 . \qquad (6.5)$$

The probability distribution of the signal points is required in (6.5), i.e., if for instance source-optimized quantizer levels (which are known to be *not* uniformly distributed) are mapped onto a signal constellation and a uniform distribution is assumed for the channel simulation, the actually used average channel-symbol energy does not equal E_s and we have to correct the true value of E_s/N_0 (i.e., the value that is used in a performance plot) appropriately. Notice, however, that for M-PSK $|s_{\gamma(i)}| = 1 \; \forall i$, so (6.5) is fulfilled independently of the probability distribution of the quantizer code-vectors. If the signal points are uniformly distributed, i.e., $\mathrm{P}(i) = \frac{1}{M} \; \forall i$, half of the symbol energy E_s is used in both the inphase and the quadrature component of the channel (if, as usual, the signal constellation is symmetric with zero mean).

6.2.1 *Conventional Hard-Decision Receiver*

At the receiver, the detector searches for the point from the modulation signal set that is the nearest neighbor of the received signal point $\tilde{z}$. This, basically, comes down to a quantization of the received signal space onto the fixed modulation signal set, again using the mean-squared-error distortion criterion. The number $\tilde{j}$ of the signal point is the output of the detector and after the inverse mapping $\tilde{i} = \gamma^{-1}(\tilde{j})$ the code-vector $y_{\tilde{i}}$ is taken from

---

[3] As usual, we implicitly assume Nyquist signaling, coherent detection, and the use of a matched filter at the receiver. For details, see Section A.3.3.

the codebook $\mathcal{Y}$ and is used as the decoded signal, i.e., $\tilde{x} = y_{\tilde{\imath}}$. Due to mathematical tractability, this simple decoding scheme is chosen for the derivation of the new SAM algorithm in the sections below.

6.2.2 *Conventional Soft-Decision Receiver*

In terms of minimum mean-squared error in the output signal $\tilde{x}$, the hard decision decoder is not best for the given transmitter. A better, soft-decision decoder, which is optimal for uncorrelated sources and memoryless channels, can be found by evaluation of (2.14):

$$\tilde{x}^{\circledast}(\tilde{z}) = \mathrm{E}_{X|\tilde{z}}\{X|\tilde{z}\} = \int\limits_{\mathcal{X}} x \cdot p(x \mid \tilde{z})dx = \frac{1}{p(\tilde{z})} \int\limits_{\mathcal{X}} x \cdot p(\tilde{z} \mid x)p(x)dx \ . \quad (6.6)$$

As the input vector is quantized, the channel input z is the same, as long as the source vector x is located within the decision region $\mathcal{X}_i$ of a code-vector y_i. Hence, we may write

$$\tilde{x}^{\circledast}(\tilde{z}) = \frac{1}{p(\tilde{z})} \sum_{i=0}^{M-1} p\big(\tilde{z} \mid z = s_{\gamma(i)}\big) \int\limits_{\mathcal{X}_i} x \cdot p(x)dx \ . \quad (6.7)$$

Since the (source-optimized) quantizer code-vector y_i is the centroid of the partition region $\mathcal{X}_i$ it represents, we have

$$y_i = \frac{1}{\mathrm{P}(i)} \int\limits_{\mathcal{X}_i} x \cdot p(x)dx \ , \quad (6.8)$$

where

$$\mathrm{P}(i) = \int\limits_{\mathcal{X}_i} p(x)dx \quad (6.9)$$

is the probability that a source vector is located in the region $\mathcal{X}_i$, i.e., $\mathrm{P}(i)$ is the known probability of the code-vector y_i. Hence, the soft-decision decoder equals

$$\tilde{x}^{\circledast}(\tilde{z}) = \frac{\displaystyle\sum_{i=0}^{M-1} p\big(\tilde{z} \mid z = s_{\gamma(i)}\big) \cdot \mathrm{P}(i) \cdot y_i}{\displaystyle\sum_{i'=0}^{M-1} p\big(\tilde{z} \mid z = s_{\gamma(i')}\big) \cdot \mathrm{P}(i')} \ , \quad (6.10)$$

where $p(\tilde{z})$ has been expanded in the denominator. For the two-dimensional AWGN channel with the inphase components marked by "*I*" and the quadrature components marked by "Q," the channel-term is given by

$$p\Big(\underbrace{\{\tilde{z}_I, \tilde{z}_Q\}}_{\tilde{z}} \mid \underbrace{\{z_I, z_Q\}}_{z}\Big) = \frac{1}{2\pi\sigma_n^2} \cdot \exp\Big(-\frac{(\tilde{z}_I - z_I)^2}{2\sigma_n^2}\Big) \cdot \exp\Big(-\frac{(\tilde{z}_Q - z_Q)^2}{2\sigma_n^2}\Big),$$

(6.11)

with $\sigma_n^2 = \frac{N_0}{2E_s}$. Although the derivation of the SAM algorithm below will be performed for the hard decision receiver, the soft-decision decoder given by (6.10) will also be used in the simulations, as it is practically important.

6.3   Principle of Source-Adaptive Modulation (SAM)

The basic idea of SAM is illustrated by Figure 6.2. As an example, the *current* source sample x is assumed to be located nearest to the quantizer reproduction level y_1 and close to the decision boundary g_2 between the reproduction levels y_1 and y_2. This means that x lies far away from the reproduction level y_0, i.e.,

$$d(x, y_1) < d(x, y_2) << d(x, y_0)   . \qquad (6.12)$$

The quantizer decision boundaries $g_i, i = 1, ..., M-1$, follow directly from the quantizer reproduction levels and the selected distance measure. In case of scalar quantization using the mean-squared error (6.1), the quantizer decision boundaries are given as the average[4] of neighbored reproduction levels, i.e., $g_i = \frac{y_i + y_{i-1}}{2}$, $i = 1, ..., M-1$. For vector quantizers the "decision boundaries" generalize to Voronoi-regions which can only hardly be described analytically. Thus, a vector quantizer is usually represented by a codebook and a distance measure. The principle of SAM is, however, more easily explained by use of the alternative representation with the quantizer decision regions.

If the modulation signal point s_1, which is allocated to the quantizer reproduction level y_1 by the mapping[5] $\gamma(i) = i$, is transmitted and a realization n of the two-dimensional channel noise is added (illustrated by the dashed circle with s_1 in its center), the received point $\tilde{z}$ will lie within the decision regions $\tilde{Z}_0$ and $\tilde{Z}_2$ with the same probability, i.e., the large distortion $d(x, y_0)$ will be produced by the receiver with the same probability as the small distortion $d(x, y_2)$.

---

[4]The limits g_0, g_M for a source distribution with an unlimited range of values are given by $b_0 = -\infty, b_M = +\infty$.

[5]This simple mapping was chosen for clarity.

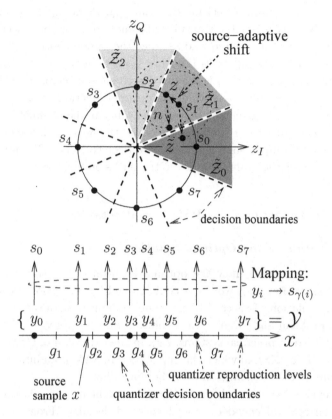

Fig. 6.2　Principle of source-adaptive modulation (SAM) (reprinted from [Goertz (2002)] with permission, ©2002 IEEE).

Now, if the transmitted signal point z is shifted away from the boundary of the decision region $\tilde{\mathcal{Z}}_0$ towards the decision region $\tilde{\mathcal{Z}}_2$, the risk to detect the signal point s_0 is reduced. Note that we perform the shift on the circle connecting the M-PSK signal points, since we have to fulfill[6] the power constraint at the channel input. On the other hand, the probability is increased to receive a point $\tilde{z}$ within the decision region $\tilde{\mathcal{Z}}_2$, which, in fact, also leads to a symbol detection error. This increase of the symbol error probability $P(\tilde{z} \in \tilde{\mathcal{Z}}_2 | z)$ leads to an increase in the source distortion (compared to a correct detection) that is the lower the closer the source

---

[6]If we had no power constraint, there wouldn't be any noisy-channel transmission problem, as we could achieve any arbitrarily small symbol error probability by a transmission power that is only large enough.

sample x lies to the quantizer decision boundary g_2. If the input sample x lies exactly on this boundary the transmitted signal point can be placed on the decision boundary between s_1 and s_2 since the same distortion $d(x, \tilde{x})$ is produced, if s_1 (i.e., $\tilde{x} = y_1$) or s_2 (i.e., $\tilde{x} = y_2$) is detected (keep in mind that the simple mapping $\gamma(i) = i$ is used to describe the principle of SAM).

Naturally, the question arises what the optimal location of the transmitted signal point z is, given the power constraint at the channel input, the modulation signal set $\mathcal{S}_M$ used for detection, the noise variance of the channel, the current input sample x, the quantizer codebook $\mathcal{Y}$, and the distance measure for the evaluation of the transmission quality.

6.4   SAM for Detection of M-PSK Signal Sets

6.4.1   *Derivation of the Optimal Solution*

In this section we describe how the optimal location of the transmitted signal point can be found, if we use a conventional hard-decision receiver for M-PSK modulation. This case allows some simplification of the general optimization problem and an analytical solution exists. Moreover, 8-PSK modulation is used in practical systems, e.g., in the extension of GSM mobile radio (EDGE-GPRS), in which the application of SAM is an interesting option for multimedia transmission over the down-link.

In the previous section we have implicitly brought up the question of how to find an optimal *encoder* for a given fixed decoder. From a more theoretical point-of-view we have discussed this already in the introduction in Section 2.4. There, we stated the following optimization problem:

$$\varepsilon^{\circledast}(\varrho) = \arg \min_{\varepsilon:\ \text{constraint}} D(\varepsilon, \varrho) , \qquad (6.13)$$

where $z = \varepsilon(x)$ is the encoder we want to optimize, $\tilde{x} = \varrho(\tilde{z})$ is the given decoder, and $D(\varepsilon, \varrho)$ is the expectation of the total system distortion that (see Section 2.4) can be written as

$$D(\varepsilon, \varrho) = \int_{\mathcal{X}} \mathrm{E}_{\tilde{Z}|x}\{d(x, \tilde{X})|x\} \cdot p(x)\, dx , \qquad (6.14)$$

with

$$\mathrm{E}_{\tilde{Z}|x}\{d(x, \tilde{X})|x\} = \int_{\tilde{Z}} d\big(x, \tilde{x} = \varrho(\tilde{z})\big) \cdot \underbrace{p(\tilde{z}\,|\,z = \varepsilon(x))}_{\text{cond. channel PDF}}\, d\tilde{z} . \qquad (6.15)$$

In the M-PSK case, the power at the channel input is the same for each channel symbol[7] and not only in an average sense, i.e.,

$$\|z\|^2 = \|\varepsilon(x)\|^2 = P \quad \forall x \,. \tag{6.16}$$

To be fair in a comparison later on, we will adopt this constraint for SAM, which means that the optimized channel input z will be placed on the circle connecting the M-PSK modulation signal points. Due to the constraint (6.16) and the fact that the distance measure $d(\cdot,\cdot)$ is strictly non-negative, the optimization problem (6.13) is turned into the problem of how to minimize (6.15) for *each* particular source vector x, i.e.,

$$z^{\circledast}(x) = \varepsilon^{\circledast}(x,\varrho) = \arg \min_{z:\|z\|^2=P} \mathrm{E}_{\tilde{Z}|x}\{d(x,\tilde{X})|x\} \,. \tag{6.17}$$

In what follows, we will abbreviate the expectation in (6.17) by

$$D(x) \doteq \mathrm{E}_{\tilde{Z}|x}\{d(x,\tilde{X})|x\} \,. \tag{6.18}$$

As we assume a hard decision receiver, the set $\tilde{\mathcal{Z}}$ of possible channel outputs is partitioned into M disjoint decision regions $\tilde{\mathcal{Z}}_j$, i.e.:

$$\tilde{\mathcal{Z}} = \bigcup_{j=0}^{M-1} \tilde{\mathcal{Z}}_j \,. \tag{6.19}$$

If a received signal point $\tilde{z}$ lies in the decision region $\tilde{\mathcal{Z}}_j$, a hard decision receiver will detect the signal point s_j, which, via the inverse index mapping $\gamma^{-1}(j)$, is decoded to $\tilde{x} = y_{\gamma^{-1}(j)}$ at the receiver output. Hence, we obtain for $D(x)$ by evaluation of (6.15):

$$D(x) = \sum_{j=0}^{M-1} \int_{\tilde{\mathcal{Z}}_j} d\big(x, \tilde{x} = \varrho(\tilde{z})\big) \cdot p(\tilde{z} \mid z = \varepsilon(x)) \, d\tilde{z} \tag{6.20}$$

$$= \sum_{j=0}^{M-1} d\big(x, y_{\gamma^{-1}(j)}\big) \int_{\tilde{\mathcal{Z}}_j} p(\tilde{z} \mid z = \varepsilon(x)) \, d\tilde{z} \tag{6.21}$$

$$= \sum_{j=0}^{M-1} d\big(x, y_{\gamma^{-1}(j)}\big) \cdot \mathrm{P}\big(\tilde{z} \in \tilde{\mathcal{Z}}_j \mid z = \varepsilon(x)\big) \,. \tag{6.22}$$

In what follows, we will make the notation shorter by omitting $\varepsilon(x)$ in (6.22), i.e., we write

$$D(x) = \sum_{j=0}^{M-1} d\big(x, y_{\gamma^{-1}(j)}\big) \cdot \mathrm{P}\big(\tilde{z} \in \tilde{\mathcal{Z}}_j \mid z\big) \,, \tag{6.23}$$

---

[7] The number of channel uses to transmit a source vector x equals $K = 1$ in our setup.

but we will keep in mind that the goal is to find the optimal transmitted signal point z for a given input source vector x. If we vary over all possible input vectors x and find an optimal signal point z in each case, we have implicitly found the optimal encoder mapping ε we are actually looking for. The quantity $P(\tilde{z} \in \tilde{\mathcal{Z}}_j \mid z)$ describes the probability that a signal point $\tilde{z}$ within the decision region $\tilde{\mathcal{Z}}_j$ is received when the signal point z is transmitted; in what follows, we will show how to compute it.

As suggested by the introductory discussion of the SAM principle, we will use the conventional transmitter as a starting point for the analysis below. We assume that the quantizer has selected the reproduction level $y_{\gamma^{-1}(1)}$ that is mapped to the modulation signal point s_1; the neighboring signal points of the M-PSK constellation are s_0 and s_2. In what follows, we will give a specialized solution for this case, but the results can be easily extended to any signal point s_j by simply replacing $0 \to j-1$, $1 \to j$, and $2 \to j+1$ in the formulas.[8]

A commonly used approximation in M-PSK systems is the assumption that the channel noise will only lead to received signal points $\tilde{z}$ that lie in the decision regions of signal points being neighbored to s_1. This is a very tight approximation [Haykin (2000)] for channels with low noise variance and it is still tight enough for our needs at higher noise variances; both facts are shown in Appendix C. Thus, (6.23) can be reformulated as follows:

$$D(x) = d(x, y_{\gamma^{-1}(0)}) \cdot P(\tilde{z} \in \tilde{\mathcal{Z}}_0 | z) + d(x, y_{\gamma^{-1}(2)}) \cdot P(\tilde{z} \in \tilde{\mathcal{Z}}_2 | z) +$$
$$d(x, y_{\gamma^{-1}(1)}) \cdot P(\tilde{z} \in \tilde{\mathcal{Z}}_1 | z) . \quad (6.24)$$

Since we assume that the received signal $\tilde{z}$ is either in the decision region $\tilde{\mathcal{Z}}_1$ (correct detection) or in one of the regions $\tilde{\mathcal{Z}}_0$, $\tilde{\mathcal{Z}}_2$ (symbol detection error), we have

$$P(\tilde{z} \in \tilde{\mathcal{Z}}_1 | z) = \left(1 - P(\tilde{z} \in \tilde{\mathcal{Z}}_0 | z) - P(\tilde{z} \in \tilde{\mathcal{Z}}_2 | z)\right) . \quad (6.25)$$

Therefore, we only have to compute the probabilities $P\{\tilde{z} \in \tilde{\mathcal{Z}}_0 | z\}$ and $P\{\tilde{z} \in \tilde{\mathcal{Z}}_2 | z\}$. For this, we introduce some definitions in Figure 6.3, which shows the upper right part of Figure 6.2.

As stated above, we seek to find a location of the transmitted signal point z that minimizes the mean-squared error in the receiver output signal

---

[8]Note that for $j = 0$ we have to set $j - 1$ to M and that for $j = M - 1$ we have to set $j + 1$ to 0, i.e., we have to perform a "wrap-around" on the PSK-circle. A general solution for any j would require to consider these special cases, which would make the notation unnecessarily complex.

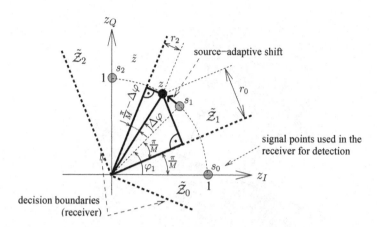

Fig. 6.3 Analysis of source-adaptive modulation for detection of M-PSK signal sets (reprinted from [Goertz (2002)] with permission, ©2002 IEEE).

$\tilde{x}$, but we restrict the point z to lie on the circle that connects the M-PSK signal points $s_0, ..., s_{M-1}$, so the energy E_s per channel symbol is not changed by SAM. Then, the location of the signal point z can be expressed by the phase difference $\Delta\varphi$ (this includes the sign!) from the M-PSK modulation signal point s_1 that is associated with the code-vector $y_{\gamma^{-1}(1)}$ selected by the quantizer.

The distances of the transmitted point z to the decision boundaries of the regions $\tilde{Z}_0$ and $\tilde{Z}_2$ are given by

$$r_0 = \sin\left(\frac{\pi}{M} + \Delta\varphi\right) \tag{6.26}$$

and

$$r_2 = \sin\left(\frac{\pi}{M} - \Delta\varphi\right), \tag{6.27}$$

respectively. The signal point s_1 can be expressed by

$$s_1 = \{\cos(\varphi_1), \sin(\varphi_1)\} \tag{6.28}$$

(the first term within the brackets is the inphase component, the second term is the quadrature component) and the transmitted signal point is given by

$$z = \{\cos(\varphi_1 + \Delta\varphi), \sin(\varphi_1 + \Delta\varphi)\} . \tag{6.29}$$

The components of the two-dimensional channel noise are independent and Gaussian. Thus, the statistical properties of the noise are the same

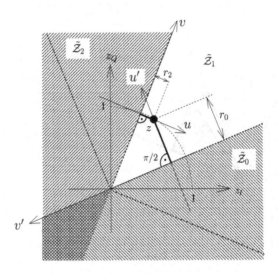

Fig. 6.4   Simplification of the integration regions (reprinted from [Goertz (2002)] with permission, ©2002 IEEE).

if we use any other set of orthogonal axes (rotational invariance of the noise [Haykin (2000)]), e.g., the axes u, v in Figure 6.4. Hence, with the definitions (6.26) and (6.27), we can express the two probabilities that are required to compute (6.24) as follows:

$$P\left(\tilde{z} \in \tilde{\mathcal{Z}}_2 | z\right) = \iint_{\tilde{\mathcal{Z}}_2} \frac{1}{2\pi\sigma_n^2} e^{-\frac{1}{2\sigma_n^2}v^2} \cdot e^{-\frac{1}{2\sigma_n^2}(u-r_2)^2} \, du \, dv \qquad (6.30)$$

and, similarly,

$$P\left(\tilde{z} \in \tilde{\mathcal{Z}}_0 | z\right) = \iint_{\tilde{\mathcal{Z}}_0} \frac{1}{2\pi\sigma_n^2} e^{-\frac{1}{2\sigma_n^2}v'^2} \cdot e^{-\frac{1}{2\sigma_n^2}(u'-r_0)^2} \, du' \, dv' . \qquad (6.31)$$

The evaluation of the integrals (6.30), (6.31) is still rather complicated. They can be simplified by following the common practice, to change the integration regions [Haykin (2000); Foschini *et al.* (1974)]: instead of the integrations over the decision regions $\tilde{\mathcal{Z}}_0$ or $\tilde{\mathcal{Z}}_2$, the integrations are carried out over the hatched half-planes in Figure 6.4, which, for instance, results in the new integration region $\hat{\tilde{\mathcal{Z}}}_2 = \{u \in (-\infty, 0), \; v \in (-\infty, +\infty)\}$. The "cross-hatched" region in Figure 6.4 is "counted" twice using this concept, but the true values of the integrals are *not* significantly changed by this approximation [Foschini *et al.* (1974)], since the values of the e^{-x^2}-functions involved in the probability density functions have very small magnitudes in the "cross-hatched" region. This is shown in Appendix C by a comparison of a simulation and theoretical results that use this approximation.

The simplification of the integration regions in (6.30) leads to

$$
P(\tilde{z} \in \tilde{\mathcal{Z}}_2 | z) = \underbrace{\int_{-\infty}^{+\infty} \frac{e^{-\frac{1}{2\sigma_n^2} v^2}}{\sqrt{2\pi\sigma_n^2}} \, dv}_{=1} \cdot \int_{-\infty}^{0} \frac{e^{-\frac{1}{2\sigma_n^2}(u-r_2)^2}}{\sqrt{2\pi\sigma_n^2}} \, du \,, \tag{6.32}
$$

which, by the substitution $t = -\dfrac{u - r_2}{\sqrt{2}\sigma_n}$, becomes

$$
P(\tilde{z} \in \tilde{\mathcal{Z}}_2 | z) = \frac{1}{2} \cdot \frac{2}{\sqrt{\pi}} \int_{r_2/(\sqrt{2}\sigma_n)}^{+\infty} e^{-t^2} \, dt = \frac{1}{2} \operatorname{erfc}\left(\frac{r_2}{\sqrt{2}\sigma_n}\right). \tag{6.33}
$$

Similarly, we obtain

$$
P(\tilde{z} \in \tilde{\mathcal{Z}}_0 | z) = \frac{1}{2} \operatorname{erfc}\left(\frac{r_0}{\sqrt{2}\sigma_n}\right). \tag{6.34}
$$

By insertion of (6.27) into (6.33) and (6.26) into (6.34) we find

$$
P(\tilde{z} \in \tilde{\mathcal{Z}}_2 | z) = \frac{1}{2} \operatorname{erfc}\left(\frac{\sin\left(\frac{\pi}{M} - \Delta\varphi\right)}{\sqrt{2}\sigma_n}\right) \tag{6.35}
$$

and

$$
P(\tilde{z} \in \tilde{\mathcal{Z}}_0 | z) = \frac{1}{2} \operatorname{erfc}\left(\frac{\sin\left(\frac{\pi}{M} + \Delta\varphi\right)}{\sqrt{2}\sigma_n}\right), \tag{6.36}
$$

respectively. The expected distortion (6.24) can now be rewritten as

$$
D(x) = d(x, y_{\gamma^{-1}(0)}) \cdot p_s'(\Delta\varphi) + d(x, y_{\gamma^{-1}(2)}) \cdot p_s'(-\Delta\varphi) +
$$
$$
d(x, y_{\gamma^{-1}(1)}) \cdot \left(1 - p_s'(\Delta\varphi) - p_s'(-\Delta\varphi)\right), \tag{6.37}
$$

using (6.25), inserting (6.35), (6.36), and by definition of

$$
p_s'(\Delta\varphi) \doteq \frac{1}{2} \operatorname{erfc}\left(\sqrt{\frac{E_s}{N_0}} \sin\left(\frac{\pi}{M} + \Delta\varphi\right)\right) \tag{6.38}
$$

with $\sigma_n^2 = N_0/(2E_s)$. The analytical result for the probability of a symbol error in hard detected conventional M-PSK (without SAM) is known [Haykin (2000)] to equal $p_s = \operatorname{erfc}\left(\sqrt{\frac{E_s}{N_0}} \sin\left(\frac{\pi}{M}\right)\right)$. It is a good cross-check of our result that this formula can also be derived from (6.35) and (6.36): the sum of both probabilities gives the probability of a symbol error if s_1 (or any other signal point) is transmitted. If we set $\Delta\varphi$ to zero (no SAM) we obtain $p_s = 2 \cdot p'_s(\Delta\varphi' = 0)$, which by use of (6.38) gives the known result.

Due to the constraint that z shall be located on the circle connecting the M-PSK signal points, we have to vary only the parameter $\Delta\varphi$ to minimize the expected distortion $D(x)$ by a source-adaptive shift.

It is necessary to discuss the range of possible values for $\Delta\varphi$: if it is limited such that the transmitted signal point z still lies in the decision region $\tilde{Z}_1$, then, no matter what the assumption for the channel noise variance is at the transmitter, the correct signal point will be detected, if the true channel has a very low noise variance. This limits the range of "allowed" values to

$$-\frac{\pi}{M} < \Delta\varphi < \frac{\pi}{M} \ . \tag{6.39}$$

Hence, the "clean-channel performance" will not be reduced by SAM, even if the noise variance is not known exactly at the transmitter.

The limitation of $\Delta\varphi$ according to (6.39) has a second justification: in (6.24) the assumption was made that, in case of a symbol error, the channel noise will only lead to received points $\tilde{z}$ that are located in the neighbored decision regions of the signal point s_1, which was selected by the quantizer (via the mapping $j = \gamma(i)$). This assumption could be violated if we allowed the transmitted signal point z to move into neighboring decision regions and, thus, the assumptions we made in the derivation of (6.37) could possibly be not fulfilled.

The optimal shift-angle can be found by standard variational techniques: in Appendix D it is shown that setting the derivative of $D(x)$ for $\Delta\varphi$ to zero results in

$$K_1 = \frac{\cos\left(\frac{\pi}{M} - \Delta\varphi^{\circledast}\right)}{\cos\left(\frac{\pi}{M} + \Delta\varphi^{\circledast}\right)} \cdot e^{K_2 \sin(2\Delta\varphi^{\circledast})} \tag{6.40}$$

with

$$K_1 = \frac{d(x, y_{\gamma^{-1}(0)}) - d(x, y_{\gamma^{-1}(1)})}{d(x, y_{\gamma^{-1}(2)}) - d(x, y_{\gamma^{-1}(1)})} \tag{6.41}$$

and

$$K_2 = \frac{1}{2\sigma_n^2} \sin\left(\frac{2\pi}{M}\right) = \frac{E_s}{N_0} \sin\left(\frac{2\pi}{M}\right) \ . \tag{6.42}$$

The optimal location of the transmitted signal point is given by inserting the source-dependent optimal shift-angle $\Delta\varphi^{\circledast}$ into (6.29).

We will discuss the solution below. Since the quantizer selects the nearest neighbor from the codebook to represent the input value (or vector)

x, $d(x, y_{\gamma^{-1}(0)}) > d(x, y_{\gamma^{-1}(1)})$ and $d(x, y_{\gamma^{-1}(2)}) > d(x, y_{\gamma^{-1}(1)})$ hold, so considering (6.41) we obtain

$$K_1 > 0. \tag{6.43}$$

The constant K_2 depends on the channel quality, for which the range

$$0\,\text{dB} < 10 \log_{10} \left(E_s/N_0 \right) < 20\,\text{dB} \tag{6.44}$$

is of main practical interest. If, additionally, we assume $M = 4$ as the lowest and $M = 1024$ as the highest possible number of M-PSK signal points, then

$$0.001 < K_2 < 100. \tag{6.45}$$

The function of $\Delta\varphi$ on the right-hand side of (6.40) is monotonically increasing within the interval given by (6.39), i.e., *if* there is a value $\Delta\varphi^{\circledast}$ that solves (6.40), there is always only one, which is the location of the minimum of (6.37). Unfortunately, (6.40) does not always have a solution within the range of allowed values, i.e., the edges $\Delta\varphi = \pm\frac{\pi}{M}$ of the allowed interval (6.39) have to be tested, if no value for $\Delta\varphi$ can be found that solves (6.40). Another major drawback is that (6.40) cannot be solved analytically for $\Delta\varphi^{\circledast}$, i.e., a zero-finding method is required, which is rather unattractive from a practical point-of-view.

6.4.2 *Test-Point Method*

A simple solution for this dilemma is to compute the distortion $D(x)$ by (6.37) for a number of test-points and select the angle $\Delta\varphi$ which gives the lowest distortion. We tried this method as illustrated by Figure 6.5. With the fixed signal point from the M-PSK modulation signal set in the center (the point $s_{1,0}$ in Figure 6.5), we placed the odd number $2L + 1$ of equally spaced test points $s_{1,l}$, $l = \pm1, \pm2, \dots \pm L$, on the circle in each decision region. This translates to the discrete steps

$$\Delta\varphi(l) = \frac{2\pi}{M(2L + 1)} \cdot l \,, \quad l = 0, \pm1, \pm2 \dots, \pm L \tag{6.46}$$

for the shift-angle. The choice of an odd number of test points avoids that one of them is located exactly on a decision boundary, so we make sure that (6.39) is never violated. In various simulations we tried several values for L. We found that a relatively small number, e.g., $L = 10$ for 8-PSK, is enough, as for $L = 20, 100, 500$ we observed (almost) no improvement.

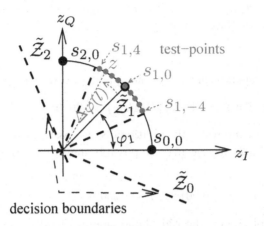

Fig. 6.5 Test-point method to find the optimal shift-angle $\Delta\varphi^{\circledast}$.

Table 6.1 Values of the factors, α, for the analytical approximation in (6.47).

M	8	16	32	64	128
α	0.47	0.205	0.0992	0.04922	0.02456

6.4.3 *Analytical Approximation*

The test-point method still requires some computational load, as we have to evaluate (6.37) $(2L+1)$ times. If that is still to much for an implementation, we can also use an analytical approximation in (6.40). The latter approach can, e.g., be realized by the substitution

$$\underbrace{\frac{\cos\left(\frac{\pi}{M} - \Delta\varphi\right)}{\cos\left(\frac{\pi}{M} + \Delta\varphi\right)}}_{\doteq g(\Delta\varphi)} \approx \underbrace{e^{\alpha \cdot \sin(2\Delta\varphi)}}_{\doteq h(\Delta\varphi)} . \tag{6.47}$$

The choice of this approximation is purely driven by the idea to analytically simplify (6.40), so it can be solved for $\Delta\varphi$. The factor α has been found empirically; some numerical results are given in Table 6.1.

Figure 6.6 shows the magnitude of the relative error

$$r(\Delta\varphi) = \left| \frac{h(\Delta\varphi) - g(\Delta\varphi)}{g(\Delta\varphi)} \right| \tag{6.48}$$

of the approximation (6.47) with the α-values from Table 6.1. We observe

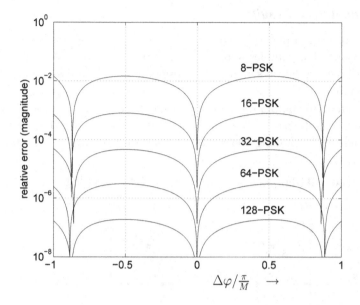

Fig. 6.6   Magnitude $r(\Delta\varphi)$ of the relative error of the approximation (6.47).

that in all cases the error (6.48) is lower than 2% and that it strongly decreases with the size M of the signal set.

Using the approximation (6.47), we obtain from (6.40)

$$K_1 = e^{(\alpha + K_2) \cdot \sin(2\Delta\varphi^\circledast)} , \qquad (6.49)$$

which is equivalent to

$$\sin(2\Delta\varphi^\circledast) = \frac{\log(K_1)}{\alpha + K_2} . \qquad (6.50)$$

As stated above we are interested in a solution for $\Delta\varphi$ within the range given by (6.39). Hence, the optimal shift angle is given by

$$\Delta\varphi^\circledast = \begin{cases} \operatorname{sign}(\beta)\dfrac{\pi}{M}, & \text{if} \quad |\beta| \geq \sin\left(\frac{2\pi}{M}\right) \\ \dfrac{1}{2}\arcsin(\beta), & \text{if} \quad |\beta| < \sin\left(\frac{2\pi}{M}\right) \end{cases} \qquad (6.51)$$

with

$$\beta = \frac{\log(K_1)}{\alpha + K_2} . \qquad (6.52)$$

Although this approach is somewhat rude, it is extremely simple to implement and it works very well; the simulation results showed no differences compared with the test-point approach with $L = 500$. Hence, we used this method for the simulations described below.

6.4.4   *Simulation Results*

We used sequences of uncorrelated Gaussian samples as a source signal for the simulations. The samples were optimally scalar quantized by three bits and an 8-PSK modulation signal set was used for the transmission over an AWGN channel.

We selected two mappings $j = \gamma(i)$ from the quantizer reproduction levels y_i to the modulation signal points s_j, they are depicted in Figure 6.7.

In the "natural" mapping the quantizer reproduction levels, with their

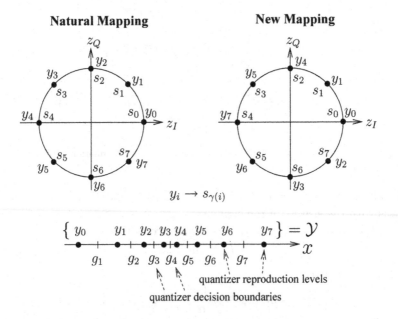

Fig. 6.7   Two mappings of the scalar quantizer reproduction y_i levels to the signal points s_j of an 8-PSK modulation signal constellation (reprinted from [Goertz (2002)] with permission, ©2002 IEEE).

natural ordering from the codebook, are wrapped around the unit circle. A large distortion is to be expected from this mapping, if the signal point s_0 is transmitted and a point in the decision region $\tilde{\mathcal{Z}}_7$ of the neighboring point s_7 is received, since the associated quantizer reproduction levels y_0, y_7 have a large distance in the source signal space.

The "new" mapping was designed to avoid this disadvantage. Instead of one area in the modulation signal space where a symbol error causes

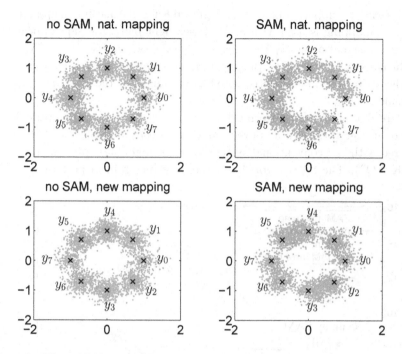

Fig. 6.8   Effect of SAM on the locations of the received signal points $\tilde{z}$ using the two mappings from Figure 6.7; $10\log_{10}(E_s/N_0) = 11$ dB.

a very large source signal distortion (as in the natural mapping), now the neighbored points from the modulation signal set are allocated to quantizer reproduction levels that lie closer in the source signal space. Although for the three pairs of signal points $s_1 \leftrightarrow s_2$, $s_3 \leftrightarrow s_4$, $s_5 \leftrightarrow s_6$ the corresponding distances in the source signal space are somewhat larger than for the "natural" mapping, the distance in the source signal space that corresponds to the signal points $s_7 \leftrightarrow s_0$ is much smaller now.

The simulations were carried out for both mappings, with and without SAM. The effect of SAM on the locations of the received points in the modulation signal space is depicted in Figure 6.8, where the quantizer levels y_i are noted beside the modulation signal points they are assigned to by the mapping. The SAM algorithm moves the transmitted signal points away from the locations, where the risk is high to get a large distortion in the source signal space due to channel noise. This produces the gap[9] between

_____

[9]Due to the use of SAM, but also because of the non-uniform probability distribution

s_0 and s_7 in the upper right plot in Figure 6.8 for the natural mapping. If the new mapping is used with SAM (lower right plot), the gaps appear between the signal points $s_1 \leftrightarrow s_2$, $s_3 \leftrightarrow s_4$, $s_5 \leftrightarrow s_6$, and $s_7 \leftrightarrow s_0$. Note that the gaps are smaller than the gap at $s_7 \leftrightarrow s_0$ for the natural mapping, since the distortions in the source signal space caused by the symbol errors are smaller.

Figure 6.9 shows the performance of an 8-PSK system with optimal scalar quantizer (SQ) and a conventional hard-decision receiver: SAM is compared with the conventional system for both mappings (plots marked "8-PSK, SQ"). The SAM algorithm shows a strong gain in performance

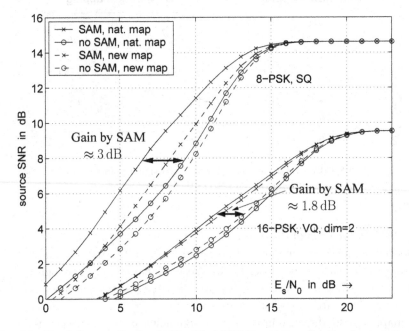

Fig. 6.9 Simulation results for SAM with 8-PSK/three-bit optimal scalar quantizer (SQ) and 16-PSK/two-dimensional four-bit vector quantizer (VQ); uncorrelated Gaussian source signal; *hard*-decision receiver (results for 8-PSK reprinted from [Goertz (2002)] with permission, ©2002 IEEE).

that is larger for the natural mapping. At $10\log_{10}(E_s/N_0) = 8$ dB, SAM

---

of the reproduction levels of a PDF-optimized quantizer, the transmitted signal points can possibly have non-zero mean, which leads to undesired peaks in the power spectral density of the continuous-time channel input. This effect can easily be avoided by a systematic rotation of the signal constellation.

achieves a gain of 3.5 dB in source SNR for the natural mapping and a gain of 1.8 dB for the new mapping, i.e., the best performance is achieved by SAM using the natural mapping. Although this is somewhat surprising, we can explain it by considering the probability distribution of the quantizer indices: although the source distortion by the "wrap-around" $s_0 \leftrightarrow s_7$ is much larger than all individual distortions in the new mapping, such a "wrap-around" can be efficiently avoided by SAM (see the large gap in the upper right plot in Figure 6.8). As the code-vectors that are allocated to the signal points s_0, s_7 have the lowest probabilities, the moderate distortion caused by SAM at these signal points is mitigated by averaging over many other source samples. The highly probable signal points, however, are assigned to quantizer levels that are direct neighbors in the source signal space; hence the distortion by symbol errors is low anyway and can also efficiently be lowered even more by SAM. Compared over both mappings, SAM achieves a gain of about 3 dB in terms of transmission power.

In Figure 6.9 we also compare a 16-PSK system with a source-optimized two-dimensional vector quantizer ($N_Y = 16 = M$); the performance curves are marked by "16-PSK, VQ, dim. 2." For the natural mapping we sorted the quantizer codebook in such a way that code-vectors which are close in the source signal space are assigned to indices that are also close in the "integer number" space.[10] Again, we observe strong gains in source SNR by SAM; the index mapping is less important than in the scalar case, as a perfect ordering of the codebook is only possible in the scalar case and it is the harder to achieve the larger the vector dimension is. Compared over both mappings, SAM achieves a gain of about 1.8 dB in terms of transmission power in the vector-quantizer case.

As yet, we have only considered conventional hard-decision receivers. In practice, however, we are interested to use soft-decision receivers to be able to exploit the reliability information we get from the channel output. We performed the same simulations as above with the only difference that we used the *conventional* soft-decision receiver described in Section 6.2 (which does *not* know about SAM we perform at the transmitter), as we aim at a simple algorithm to improve a system with transmitter-only modifications. The results are depicted in Figure 6.10. Qualitatively, we obtain very similar results as in Figure 6.9, the main difference is that in all simulations we observe gains of 1–2 dB in transmission power over the hard-decision receiver. Hence, SAM works well for both types of receivers.

---

[10]Such an ordering can be achieved by the Pseudo-Gray coding algorithm [Zeger and Gersho (1990)].

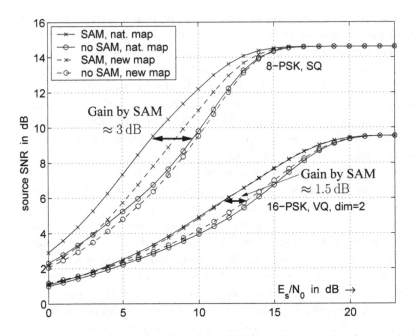

Fig. 6.10   Simulation results for SAM with 8-PSK/three-bit optimal scalar quantizer (SQ) and 16-PSK/two-dimensional four-bit vector quantizer (VQ); uncorrelated Gaussian source signal; *soft*-decision receiver.

At this point, it is interesting to recall the discussion in Sections 2.3, 2.4. The conventional soft-decision receiver is optimal (in the minimum-mean-squared-error-sense) for the transmitter of the system model in Section 6.2. In the simulations in Figure 6.10 we have used this receiver but a modified SAM-transmitter and we achieved strong quality gains.

One could ask, why we did *not* use a conventional soft-decision receiver in the derivation of SAM, because most of the practically used receivers will use reliabilities at the channel outputs whenever they are available. The reason is simply that the derivation of SAM in this case becomes intractable, because the decoder given by (6.10) would have to be used in (6.15) which would make it impossible to separate integrals over decision regions (as in (6.23)) that can be solved analytically.

In Figure 6.11 we, finally, compare the results for the soft-decision receiver and the natural mapping from Figure 6.10 with the information-theoretical performance limits. The channel capacity for the two-dimensional AWGN channel with discrete PSK input alphabet is dis-

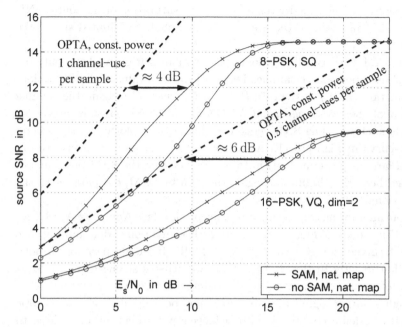

Fig. 6.11 Comparison of the SAM performance results with the information-theoretical limits.

cussed in Appendix A.3.3; the curves in Figure A.14 indicate that for $10 \log_{10}(E_s/N_0) < 25$ dB there is no capacity-gain by use of more than 64 PSK signal points. Hence, we can use this curve to find the optimal performance theoretically attainable (OPTA) for a Gaussian source that is transmitted over a two-dimensional AWGN channel at 0 dB $< 10 \log_{10}(E_s/N_0) < 23$ dB and with constant power for *each* channel-use. As we have one channel-use per source sample for scalar quantization, the curve for $M = 64$ from Figure A.17 represents the theoretical performance limit for this case. Note that we use all $(M = 64)$ possible locations on the unit-circle as the (quasi) continuous channel input alphabet for our OPTA curves, since SAM is also allowed to use any point on the circle.

We observe a gap of about 4 dB to the performance limit (in the slope of the performance curve). The reason is that capacity achieving schemes usually require infinitely large block lengths and complex source and channel coding algorithms. Both is not fulfilled, as the SAM scheme is very simple and the block-length is only one for scalar quantization. Hence, the gain by SAM that, in the slope of the curve, almost makes half the way to the theoretical limit is quite remarkable. The saturation of the source SNR for high E_s/N_0 is due to the use of a signal alphabet of limited size at the receiver.

For SAM with two-dimensional vector quantization we have only 0.5 channel-uses per source-sample, i.e., we have to evaluate the distortion-rate function (DRF) at the rate $R = 0.5 \cdot C$, where C is again the channel capacity of the two-dimensional AWGN channel with constant power for *each* channel-use. The SNR-version of this DRF is plotted as the second OPTA-curve in Figure 6.11. As above and for the same reasons, we observe a large gap (about 6 dB in this case) between the SAM performance and the theoretical limits, but again the gain by SAM, compared with the conventional system, is remarkable.

From an information theoretical point-of-view (see Appendix A.4.3, Figure A.17), the use of more than 8 or 16 PSK signal points does not improve the performance for $10 \log_{10}(E_s/N_0) < 10$ dB. It should be noticed, however, that exactly in this E_s/N_0-region we achieve strong gains by SAM by use of more (transmitted) signal points. Hence, we can deduce that although a larger signal alphabet may not be helpful in an asymptotic sense, it can definitely be helpful to achieve a performance that is closer to the theoretical limits with systems that must fulfill constraints on delay or complexity or both.

6.5   SAM for Quadrature Amplitude Modulation

In the previous section we derived SAM for receivers that use an M-PSK signal set for detection. We showed that in the M-PSK case the power constraint (6.16) is easy to fulfill, if the transmitted signal points are restricted to be located on the circle connecting the M-PSK signal points. In what follows we show that SAM can also be applied to the more general QAM signal constellations.

6.5.1   *Discussion of Potential Signal-Point Locations*

The first idea could be, to allow only those locations for an optimization by SAM that have the same power as the signal point from the fixed QAM signal set that has been selected by the quantizer. This way, the energy would not be changed by SAM and the average power constraint would be automatically fulfilled (as long as (6.5) is considered in the design of the conventional system).

Figure 6.12 illustrates this approach. The dashed circles indicate areas

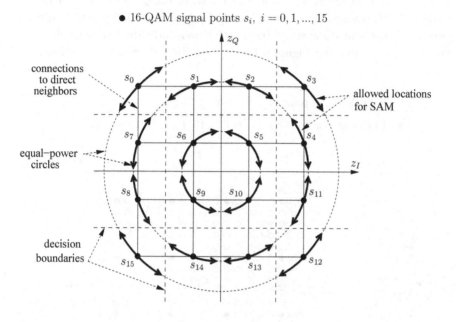

- 16-QAM signal points s_i, $i = 0, 1, ..., 15$

Fig. 6.12   Location of the transmitted signal points for 16-QAM without change of the transmission power of each transmitted signal point.

of constant transmission power. As above for M-PSK, we will restrict the location of any transmitted signal point such that it remains within the decision region of the signal point that is addressed by the quantizer. These "allowed" regions are indicated in Figure 6.12 by the bold circle-segments. The reason for this limitation is that the detection at the receiver will "always" be correct if the noise on the channel is very low, even if the noise variance is not exactly known at the transmitter. Moreover, it is reasonable, as in the analysis for QAM (as for M-PSK) the assumption is made that symbol errors will only lead to detected signal points that are directly neighbored to the correct signal point. This assumption would be violated if the transmitted signal points were allowed to move anywhere on their equal-power circles.

The problem with the allowed locations of the SAM-optimized signal points as given by Figure 6.12 is that the possible variations are rather limited and, at the same time, the signal points have more than two direct neighbors. It is, e.g., impossible for a signal point in the decision region of s_5 to move away from s_2 *and* s_4 at the same time. To get more flexibility, we can exploit that QAM does *not* have constant energy for each symbol anyway. Hence, if we want to apply the SAM principle, we are free to fulfill the power constraint on average and not for each individual signal point. This idea is illustrated by Figure 6.13. As described above for M-PSK we

- 16-QAM signal points $s_{i,0}$, $i = 0, 1, ..., 15$

- test points for SAM $s_{i,l}$, $i = 0, 1, ..., 15$, $l = -12, ..., 0, ..., 12$

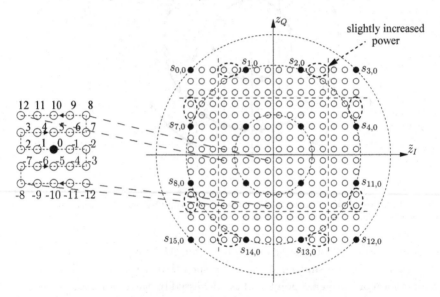

Fig. 6.13   Locations of the test-points for SAM with 16-QAM in 5 × 5 matrices.

use the test-point method. The distortion $D(x)$ for each test-point can be computed by similar approximations as in the PSK-case, but the notation is much more involved as several cases have to be distinguished: if the signal points in the corners of the constellation are selected by the quantizer, only two direct neighbors have to be considered in the computations of the probability of a symbol error. The points on the edges have three neighbors and only the four central points have four direct neighbors each. In all of these cases, the computation of the symbol error probability is straightforward,[11] as the decision regions are parallel to the axes z_I and z_Q. Hence, we will omit the details of the distortion function here. The locations of the test-points in Figure 6.13 need, however, some discussion.

The points in the corners are not allowed to move to locations with higher energy. The reason is that the average power constraint in principle cannot be fulfilled by SAM at *each* time instant, so the SAM algorithm would always select the highest energy possible for each of these points, as then the risk for a symbol error and, thereby, for large distortion in the decoder output signal would be minimal. As later on we renormalize[12] the average value for E_s/N_0 that has really been used in the simulation, it doesn't make sense to systematically move the points towards higher energy, as this increase is compensated anyway.

As a result of our restriction, the corner points will only hardly be moved at all, as the slight increase in distance to one neighboring point will cause a strong decrease of the distance to the other. With similar justification we have placed the test-points for the signal points on the edges of the constellation (only two points cause a slight increase in transmission power). Only the four central points have a full 5×5 matrix of test-points which are checked by the SAM algorithm.

From a practical point-of-view, the restrictions on the locations of the test-points described above have another advantage: the peak power used by SAM is limited to exactly the same value as for the conventional QAM transmitter. Therefore, the requirements on the dynamical range of the power amplifiers are not stronger if SAM is used.

During a simulation we have to measure the truly used average transmission power P_n'. Since we assume that the normalized channel model (see Appendix A.3.3.4, Figure A.12) with a noise-variance of $\sigma_n^2 = \frac{N_0}{2E_s}$ is used, the average power P_n of the signal constellation without SAM is normalized to one according to (6.5). Hence, after the simulation for one value of E_s/N_0 the location of the resulting source SNR-value must be shifted by $10 \log_{10}(P_n')$ dB on the E_s/N_0-axis (this includes a sign!) towards higher values of E_s/N_0.

---

[11] and easier than in the M-PSK case

[12] Such a renormalization is also required because of the non-uniform probability distribution of the quantizer levels.

6.5.2 *Simulation Results for SAM with QAM*

For a simulation we again selected a Gaussian source signal that was optimally scalar quantized by 4 bits. We mapped the quantizer indices to a standard 16-QAM constellation, which we used for detection at the conventional (soft and hard decision) receiver, and we implemented SAM for QAM by the test-point method as described above.

The received signal points at $10\log_{10}(E_s/N_0) = 12\,\mathrm{dB}$ are depicted in the left plot of Figure 6.14 for the conventional system. In the right plot we show the received signal points if SAM is used at the transmitter; in this case 0.027 dB more transmission power were required.[13]

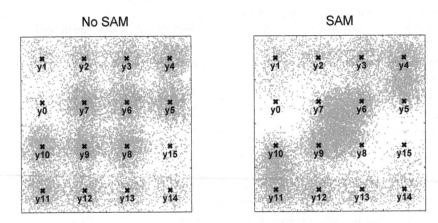

Fig. 6.14 Effect of SAM on the locations of the received signal points $\tilde{z}$ using a 16-QAM signal constellation at $10\log_{10}(E_s/N_0) = 12\,\mathrm{dB}$ (SAM requires 0.027 dB more transmission power than the conventional system). The mapping from the code-vectors y_j to the signal points is indicated. The quantizer codebook is sorted according to $y_0 < y_1 < \ldots < y_{15}$.

Similar as for M-PSK, Figure 6.14 shows that SAM moves the transmitted signal points away from locations, where the risk is high to get large distortions in the source signal space due to channel noise. It is obvious that the additional flexibility we get from fulfilling the power constraint only "on average" (by the selection of the test-points as proposed in Figure 6.13) is really needed, as, e.g., the area around the origin is populated by a

---

[13]Over the whole range of E_s/N_0, SAM required more transmission power than the conventional system, but the maximum increase, which occurs at low E_s/N_0, was smaller than 0.3 dB.

lot of realizations. Signal points in this area can only be selected by SAM
if the individual power of each symbol is allowed to change.

Figure 6.15 shows the performance of SAM (including the power renormalization described above), for the code-vector mapping that is depicted in Figure 6.14. As in the M-PSK case we achieve large gains by

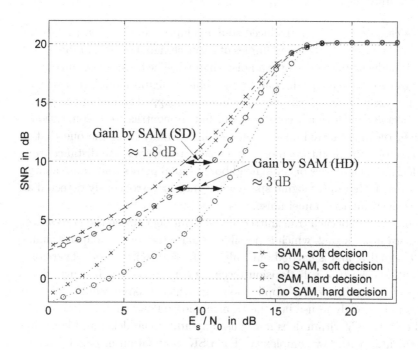

Fig. 6.15 Performance of SAM for 16-QAM signal constellations.

SAM for hard and soft-decision receivers; the mapping of the code-vectors
to signal points has, however, strong influence on the total performance
of the system. Among several mappings we tested, the one in Figure 6.14
performed best. One idea for its construction is to place the quantizer-
levels with high probability in the center of the constellation, so only a
small amount of energy is "consumed" by these points. The second design-
condition is to achieve large distances in the modulation signal space between those quantizer-levels that have large distances in the source signal
space. The third condition is to achieve a signal constellation with zero
mean. For large signal constellations the mapping could possibly be optimized by a numerical approach such as "binary switching" [Zeger and

Gersho (1990)] (see also Chapters 3, 5). However, even if a "bad" mapping is chosen or must be used due to system constraints, SAM achieves strong gains over the conventional QAM scheme.

6.6   Conclusions

In this chapter we have introduced source-adaptive modulation (SAM) for improved transmission of the samples of a source signal or the parameters of a multimedia source codec over a noisy channel. The SAM algorithm uses a quantizer for source encoding and a non-binary digital modulation scheme for the transmission. Each quantizer index is mapped to one of the points from a modulation signal constellation, but, in contrast to a conventional scheme, the transmitted signal point is moved away from the one that is addressed by the quantizer index: the goal is to increase the distance from the decision boundaries of those neighbored signal points that cause a large distortion in the source signal space when they are erroneously detected at the receiver due to channel noise.

The location of each transmitted signal point depends on the *current un-quantized* input signal, which is possible because the source-adaptive choice of the signal points is carried out only at the transmitter side. Therefore, conventional hard or soft decision receivers can be used and, hence, SAM is especially attractive for applications, in which a large number of such receivers is already in use but the transmission quality shall be improved.

The SAM algorithm does not cause any additional delay and it can be implemented with low complexity. For PSK modulation in particular, we stated an analytical approximation of the optimal solution that can be realized with almost no increase in complexity compared with the conventional system.

The simulation results for 8-PSK and 16-QAM modulation constellations show that SAM achieves strong gains in source SNR (or in transmission power) over a conventional system with both hard and soft decision receivers.

Chapter 7

Source-Adaptive Power Allocation*

7.1 Introduction

Similar as in Chapter 6, we consider the problem of transmitting continuous-valued source samples over a noisy channel by a source-adaptive modulation scheme. Due to a constraint on the tolerable delay, joint encoding of long blocks of source samples is assumed to be infeasible. As further constraints on the energy consumption usually exist, the tolerable transmission power and the computational complexity may be strongly limited, the latter especially for the decoders in a broadcast scenario. Another example is the transmission of multimedia data over the down-link of a mobile radio network: if the channel quality is fairly good, only a small number of data bits may be error-protected by a channel code. Thus, the continuous-valued parameters of an advanced source codec (e.g., gain factors or filter coefficients) can be viewed as the source signals in our setup that (at least in part) are transmitted without forward error protection.

In what follows, our goal is to achieve quality improvements by transmitter-only modifications, while conventional, low-complexity receivers are used. This is an attractive feature if many receivers (e.g., cell phones in the example above) are already in use. The basic idea of the new algorithm is to adjust the power for the transmission of the individual bits of a quantizer index in such a way that, with a fixed average energy per source sample, the reconstruction quality at the decoder output is maximized. The bit transmission power is adapted to the current *unquantized* source sample and not only to the average sensitivity of a particular bit

---

*Based on "Source-Adaptive Power Allocation for Digital Modulation" by N. Goertz and E. Bresch, which appeared in *IEEE Communications Letters*, vol. 7, no. 12, pp. 569–571, December 2003. ©2003 IEEE.

position of the quantizer index as, for instance, in unequal error protection. Although the idea of source-adaptive power allocation (SAP) is similar to source-adaptive modulation (SAM) stated in the previous chapter, we introduce SAP in a separate stand-alone chapter as the optimization algorithm is different from SAM. In Section 7.2 we describe a specialized system model and in Section 7.3 we introduce the theoretical background of SAP. In Section 7.4 we briefly describe the conventional soft-decision receiver, and in Section 7.5 we provide simulation results.

7.2   System Model

The model of our communication system is depicted in Figure 7.1. The

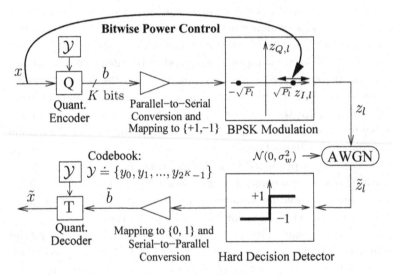

Fig. 7.1 Model of a communication system for the transmission of continuous-valued source samples x.

discrete-time continuous-valued source samples x are fed into a K-bit (scalar or vector) quantizer, which outputs the bit vector[1]

$$b \doteq \{b_1, b_2, ... b_K\} . \tag{7.1}$$

---

[1]We omit the time index, as the SAP algorithm works independently for each source vector.

The bits $b_l \in \{0,1\}$, $l = 1,...,K$, are serially transmitted using BPSK or QPSK[2] modulation. If QPSK is used with Gray-coded bit labels, the detection at the receiver can be carried out independently for the inphase and the quadrature component, i.e., the problem reduces to two independent BPSK-detections. Therefore, and for clarity, we restrict the following discussion to BPSK modulation.

The K-bit vector b is transmitted by K channel uses. In contrast to a conventional setup, each channel symbol z_l, $l = 1,...,K$, is sent with an *individual* amplitude[3] $|z_l| = \sqrt{P_l}$, where P_l is the transmission power of the channel-symbol z_l. The total energy E equals the sum over all energies $E_l = P_l \cdot T_s$ of the individual bits (channel-uses), i.e.

$$E \doteq \sum_{l=1}^{K} E_l = \sum_{l=1}^{K} P_l \cdot T_s = T_s \cdot \sum_{l=1}^{K} |z_l|^2 , \qquad (7.2)$$

where T_s is the bit-symbol transmission period in a continuous-time system model with Nyquist signaling and matched filter (see Appendix A.3.3). Note that by (7.2) the peak power for a channel-use is strictly limited to $P_l \leq E/T_s$.

The transmission channel is assumed to introduce additive white Gaussian noise (AWGN) with a constant variance $\sigma_w^2 = N_0/(2T_s)$ at the sampled matched filter output, where $N_0/2$ is the noise power spectral density on the continuous-time channel. Hence, the conditional probability density function for the Gaussian channel is given by

$$p_c(\tilde{z}_l \mid z_l) = \frac{1}{\sqrt{2\pi}\sigma_w} e^{-\frac{1}{2\sigma_w^2}(\tilde{z}_l - z_l)^2} . \qquad (7.3)$$

In the simplest case, the receiver takes hard decisions (HD) for the incoming symbols $\tilde{z}_l$, converts the resulting bit stream into the bit vectors $\tilde{b}$, and obtains the output samples $\tilde{x}$ by a codebook table look-up. Soft-input decoders, which are described in Section 7.4, can also be used in practice, but due to mathematical tractability we will use a HD receiver for the derivation of the power allocation algorithm in the next section.

7.3 Principle of Source-Adaptive Power Allocation

Our aim is to achieve a performance gain in terms of source SNR through the efficient use of transmission power. To illustrate the concept, we assume

---

[2]binary/quaternary phase shift keying

[3]We will use the term BPSK further on, because we still transmit only a single bit per channel use.

a scalar three-bit quantizer for the source samples and we consider the case depicted in Figure 7.2: the current *un*quantized input sample lies very close to the decision boundary between the quantization levels y_1 (Gray-mapped to the bit vector $b^{(1)} = \{0, 0, 1\}$) and y_2 (Gray-mapped to the bit vector $b^{(2)} = \{0, 1, 1\}$). As both quantization levels y_1 and y_2 have almost the

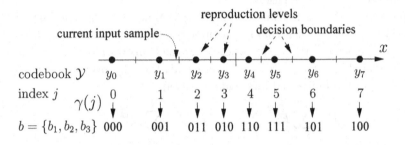

Fig. 7.2   Three-bit scalar quantizer with Gray bit-mapping.

same distance from the input sample x, it does not really matter whether the system transmits the bit vector assigned to y_1 or to y_2. Hence, one can save the transmission energy of the center bit—the only bit in which both bit vectors differ—or, if the total energy E for the bit vector is to remain constant, one may increase the energies of the first or the last bit, which results in an improved robustness against channel noise. Due to the use of the Gray labeling, the principle applies to any two neighboring quantizer levels. Similar observations can be made for vector quantizers if Gray-like labelings are used, which can be obtained, e.g., by Pseudo-Gray coding [Zeger and Gersho (1990)].

The example above raises the question of how to optimally distribute the transmission energy among the bits of a particular bit vector at the quantizer output, given the current *un*quantized input x. In order to tackle the problem mathematically, we define the conditional expected distortion

$$D(x) \doteq \mathrm{E}_{\tilde{Z}|x}\{d(x, \tilde{X})|x\}\,, \tag{7.4}$$

given a particular input source vector x, where $d(x, \tilde{x})$ is the mean-squared error between the input source vector and its reproduction at the receiver output. The philosophy is to select an optimal transmitter for a given receiver as described in Section 2.4. Our constrained is here that we want to use a fixed amount E of energy for the transmission of each particular source vector. We have shown in Section 2.4 that for this type of optimization problem we may equivalently minimize the expected distortion (7.4) instead of the total expected system distortion that is harder to compute.

As we use a hard-decision receiver according to Figure 7.1, we can re-write the expectation (7.4) as

$$D(x) = \sum_{\tilde{b} \in \mathcal{B}} d(x, y_{\gamma^{-1}(\tilde{b})}) \cdot P(\tilde{b} \mid b = Q(x)) , \qquad (7.5)$$

where $\mathcal{B}$ denotes the set of all possible K-bit vectors and $b = Q(x)$ denotes the bit-vector that is generated by the quantizer from the current source-vector x. The notation $y_{\gamma^{-1}(\tilde{b})}$ is used for the quantizer code-vector that is addressed by the bit-vector $\tilde{b}$ via the inverse bit mapping $\gamma^{-1}()$ as illustrated by Figure 7.2. Moreover, $P(\tilde{b} \mid b)$ denotes the transition probability with which a certain bit vector $\tilde{b}$ is detected, if the bit vector b was transmitted.

Since the channel noise is statistically independent, we can compute the bit-vector transition probabilities by the product

$$P(\tilde{b} \mid b) = \prod_{l=1}^{K} P(\tilde{b}_l \mid b_l) , \qquad (7.6)$$

with the bit-realizations $\tilde{b}_l, b_l \in \{0, 1\}$. The bit transition probabilities are given by

$$P(\tilde{b}_l \mid b_l) = \begin{cases} 1 - p_l, & \text{if } \tilde{b}_l = b_l \\ p_l, & \text{if } \tilde{b}_l \neq b_l \end{cases} , \qquad (7.7)$$

where the bit error probability p_l of the BPSK modulation with coherent detection and hard decisions (with matched filter) is given by (e.g., [Haykin (2000)])

$$p_l = \frac{1}{2} \operatorname{erfc}\left(\sqrt{E_l/N_0}\right) . \qquad (7.8)$$

Note that the individual bit energies $E_l = P_l \cdot T_s$ we seek to optimize appear in (7.8).

The goal of our power control is to minimize the expected distortion (7.5) for the current input x by choosing the optimum energies

$$\{E_1^{\circledast}, ..., E_K^{\circledast}\} = \arg \min_{E_1, ..., E_K: E = \sum_l E_l} D(x) \qquad (7.9)$$

for the bits, under the constraint (7.2) that the energy E used to transmit the whole bit-vector is constant.

Since an analytical solution is difficult to obtain, we solve (7.9) by sys-tematically trying out a number of different energy distributions. We allow discrete amplitudes for the channel input symbols, whereby the step size is chosen to be $\Delta = \frac{1}{5}\sqrt{E_s/T_s}$, i.e., one fifth of the average symbol amplitude, where $E_s \doteq E/K$ is the *average* bit energy. Obviously, when testing all possible amplitude combinations, it must be ensured that (7.2) is never violated. Hence, for each tested energy distribution we have to select only $K-1$ different bit energies; the last one follows from the constraint (7.2).

Corresponding to the example in Figure 7.2, Figure 7.3 shows the op-

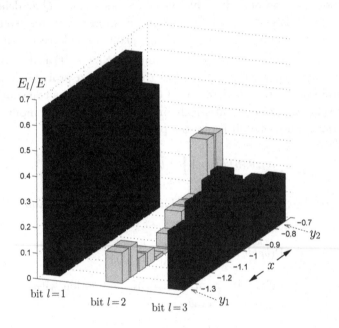

Fig. 7.3   Optimal bit-symbol energies E_l for $y_1 < x < y_2$, with the quantizer levels $y_1 = -1.34$, $y_2 = -0.76$, and the quantizer bit vectors $\gamma(1) = \{0,0,1\}$, $\gamma(2) = \{0,1,1\}$ (bit $l = 1$ is leftmost). The average "channel SNR" is $10\log_{10}(E_s/N_0) = 1.2\,\text{dB}$ with $E_s = E/K$ and E is the fixed energy for the transmission of one K-bit quantizer output bit-vector.

timum transmit energy distribution for 12 equally spaced input values x within the range $y_1 < x < y_2$, where, for scalar Lloyd-Max quantization of a Gaussian source signal with $K = 3$ bits, two of the quantizer reproduction levels are given by [Jayant and Noll (1984)] $y_1 = -1.34$ and $y_2 = -0.76$. In accordance with our previous analysis we find that, if x lies in the middle between the quantization levels, the center bit is allocated no transmission energy.

7.4   Conventional Soft-Decision Receiver

Although the optimization of the energy allocation at the transmitter is performed for a conventional HD receiver, we may use a soft-decision (SD) receiver in practice that assumes a conventional transmitter (no SAP). For simplicity and to avoid large complexity, we will assume that the source signal is uncorrelated.[4] Then, the decoder producing the minimum mean squared error in the output signal is given by (see Section 2.3)

$$\tilde{x} = \mathrm{E}_{X|\tilde{z}}\{X|\tilde{z}\} = \sum_{b \in \mathcal{B}} y_{\gamma^{-1}(b)} \cdot \mathrm{P}(b \mid \tilde{z}) \,, \qquad (7.10)$$

where $\tilde{z} \doteq \{\tilde{z}_1, ..., \tilde{z}_K\}$ contains the received channel values for all bits of a transmitted quantizer level. It is easy to show (Bayes-rule) that the bit vector a posteriori probabilities can be computed by

$$\mathrm{P}(b \mid \tilde{z}) = A \cdot \prod_{l=1}^{K} e^{-(\tilde{z}_l - z_l)^2/(2\sigma_w^2)} \cdot \mathrm{P}(b) \,, \qquad (7.11)$$

where the l-th bit $b_l \in \{0, 1\}$ of the quantizer output bit-vector b is mapped to the channel-input values $z_l = \sqrt{E_s/T_s} \cdot (1 - 2 \cdot b_l)$. In (7.11), we have used the Gaussian distribution (7.3) of the channel noise; the constant A normalizes the left-hand side such that it sums up to one over all possible bit vectors b. Note that we only use the *average* energy $E_s \doteq E/K$ per data bit, as SAP is unknown at the receiver.

7.5   Simulation Results

Figure 7.4 depicts the source SNR vs. the average channel SNR[5] $10\log_{10}(E_s/N_0)$ for scalar three-bit quantization and two-dimensional three-bit vector quantization, in both cases for a Gaussian input signal. In the VQ-case we correlated the source signal by first-order recursive low-pass filtering (coefficient $a = 0.9$, see Figure 3.2).

---

[4]If the source signal *is* correlated, the decoder could be improved without additional delay as decribed in Chapter 3 by use of all received channel values up to the current time instant in the estimation (7.10).

[5]In order to stick to general practice, we use the average energy E_s per bit (divided by the noise power spectral density N_0) as a channel-quality parameter. Hence, the source-SNR-results for scalar and vector quantization given in Figure 7.4 are not directly comparable.

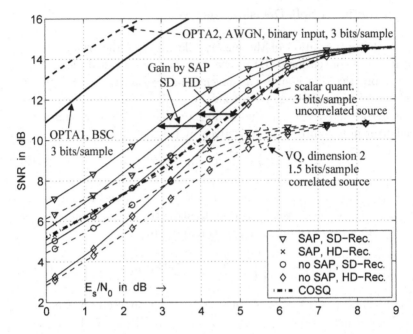

Fig. 7.4   Performance of SAP; Gaussian source signals; scalar and vector quantization; HD and SD receivers. Comparison with channel-optimized scalar quantization (COSQ) and the optimal performances theoretically attainable (OPTA).

Gains of more than 1 dB in E_s/N_0 can be achieved by SAP for both scalar and vector quantization. This is true for conventional HD receivers as well as for SD receivers.

For comparison, we have also included the performance of channel-optimized scalar quantization (COSQ) [Farvardin and Vaishampayan (1991)] on a binary symmetric channel (BSC). The latter follows from hard decisions at the channel output; the bit-error probability is given by $p_e = \frac{1}{2}\mathrm{erfc}\sqrt{E_s/N_0}$ (e.g., [Haykin (2000)]). The COSQ-scheme is clearly outperformed by SAP (curve labeled "SAP-HD," we achieve a gain of about 1 dB in E_s/N_0), but SAP requires, in contrast to COSQ, that the energy may be selected individually for each bit at the channel input.

For the uncorrelated Gaussian source we also included the optimal performances theoretically attainable (OPTA; for details, see Appendix A.4.1) in Figure 7.4, which follow from the distortion-rate function of the source, evaluated at the channel-capacity times the number (three in our example) of channel-uses per source sample. The OPTA1-curve in Figure 7.4 provides the theoretical limit for the system with the HD receiver, while the OPTA2-curve gives the same for the system with the SD receiver. Both curves are taken from Figure A.15 for $K/N = 3$ (as we have 3 channel-uses per source sample).

For the performance limits we have used the capacities of binary input channels (the BSC for the OPTA1-curve and the binary input AWGN channel for the OPTA2-curve) although the SAP algorithm does not use constant transmission power for each channel-use. The reason is that by SAP we still want to transmit one bit per channel-use, we only adjust the transmission power because of the varying sensitivities of the bits in terms of source signal quality at the decoder output. In the sense of information theory, an ideal source encoder would generate uniformly distributed, independent bits that are all equally "sensitive." Hence, there is, asymptotically, no need (and no gain from the freedom) to adjust individual bit energies.

Although we observe large gaps between the OPTA and the SAP curves it should be noticed that a strong increase in performance (compared to the conventional scheme) is achieved by SAP *without* additional algorithmic delay, i.e., we have no coding of long blocks, which is usually required by capacity-achieving schemes. Hence, the gain by SAP is quite remarkable.

7.6 Conclusions

The transmission quality of continuous-valued source signals can be significantly improved by source-adaptive power allocation (SAP); gains of more than 1 dB in transmission power are achievable. The SAP algorithm requires transmitter-only modifications, while conventional SD or HD receivers can be used which makes SAP especially attractive in situations where a relatively small number of transmitters exists and a large number of conventional receivers is already in use. Such a situation can be found, e.g., in mobile radio or in radio broadcast.

Chapter 8

Concluding Remarks

In this work we have dealt with joint source-channel coding and decoding for continuous-amplitude sources such as (the codec parameters of) speech, audio, or image signals. We discussed the topic mainly from the perspective of mobile communications, as in this field the system requirements are very challenging: we have to cope with strong interference on the transmission channels, the complexity and memory limitations for the mobile terminals are significant, and the tolerable system delay is low. The latter is especially true for the main application of mobile radio, which is (and probably will always be) speech transmission for telephony. Even without any complexity constraints, limited delay turned out to be a major reason why source and channel coding necessarily work imperfectly in the sense of information theory and, hence, practical algorithms often do not perform close to the theoretical limits. We showed that it is required to (at least partially) join source and channel coding to cope with non-ideal coding algorithms and, thereby, to achieve strong performance improvements over the classical system design based on the separation principle.

While the optimal receiver for a given transmitter is known and re-alizable approximations exist for this case (e.g., iterative source channel decoding), the design of the optimal encoder (in general, or for a given decoder) turns out to be much more involved. Therefore, we also studied some special cases, in which we could find encoders that achieve significant performance improvements.

Over the years, joint source-channel coding has been a field of varying popularity in the coding community. While its basic ideas, e.g., the use of residual source redundancies for error concealment at the receiver (which was mentioned already in Shannon's classical papers), have been widely known and used for a long time, more advanced topics have been studied

139

especially in the last decade. Some examples are channel-optimized vector quantization, optimization of quantizer-index assignments, estimation-based receivers, source-controlled channel decoding, and iterative source-channel decoding for continuous-valued sources but also for variable-length encoded text data.

Up to now, most of the work on joint source-channel coding has concentrated on the improvement of circuit switched communication systems, in which noise or interference corrupts the transmitted data. Recently, however, research has been started concerning a "cross-layer" system design in packet-based communications. It seems that this field has much in common with joint source-channel coding: both areas are based on the reasonable doubt, whether classical design concepts really make sense, which strictly separate the components (or layers) of a system. Although the details of "cross-layer" design and joint source-channel coding may be different: what, for instance, is the conceptual difference between packet transmission with priority classes and classical unequal error protection?

Appendix A

Theoretical Performance Limits

A.1 Preliminary Remarks

In this chapter we will summarize theoretical performance limits for the communication problem discussed in Chapter 2. These limits are results from information theory and they are usually non-constructive, i.e., the theory does not propose specific systems that achieve the limits in practice. Moreover, the results are often only valid for unrestricted block-lengths[1] for source and channel coding. No attempt is made to strictly prove any of the information theoretic results, as several text books exist, e.g., [Gallager (1968); Cover and Thomas (1991); McEliece (2002)] where the proofs are given in full detail.

In contrast to the joint treatment of source and channel coding, which is the topic of this book, one of the fundamental results of information theory is the separation theorem. Roughly speaking it states that for infinitely large block-lengths the encoding process can be separated into source and channel coding without loss in performance. The source encoder produces a sequence of independent and uniformly distributed bits (at a bit rate that equals the capacity of the channel) and the channel code protects the bits against channel noise. In theory, error correction is perfectly possible if the code rate of the channel code (the number of data bits divided by the number of channel code bits) is not larger than the channel capacity.

The theoretical limits are upper bounds for the performance of practical coding systems, and these bounds tend to be loose, especially if the block lengths in the real system are small, due to constraints on delay and complexity.

---

[1]It should be mentioned that infinitely large block lengths are not necessarily required in each case to achieve the theoretical limits.

In what follows, we will summarize practically important results from information theory – distortion-rate functions and channel capacities – that are relevant for our problem.

A.2   Important Distortion-Rate Functions

In rate-distortion theory the average number R of bits per source sample is quantified that is required to encode a source signal with some given distortion D (rate-distortion function, RDF). The inverse, which is often more convenient to work with from a practical perspective, is called the distortion-rate function (DRF). Most of the available results for continuous-valued sources are restricted to the squared error distortion measure; thus, we will also confine our discussion to this case.

It should be mentioned that there is no guarantee that, for limited block length, a practical system exists at all that achieves the theoretical performance limits. The results are, however, useful to know the potential improvements that may be possible.

In what follows some practically important DRFs and RDFs are summarized.

A.2.1   *Memoryless Sources*

As in [Jayant and Noll (1984)] we will consider the DRFs of the source probability density functions (PDFs) given in Table A.1, which are depicted in Figure A.1. All PDFs have zero mean ($\mu_x = 0$) and the variance $\sigma_x^2 < \infty$.

A.2.1.1   *Memoryless Gaussian Source*

The RDF for the Gaussian PDF given in Table A.1 equals [Cover and Thomas (1991)]

$$R_G(D) = \max\left\{0, \frac{1}{2}\log_2 \frac{\sigma_x^2}{D}\right\} = \begin{cases} \frac{1}{2}\log_2 \frac{\sigma_x^2}{D}, & 0 \leq D \leq \sigma_x^2 \\ 0, & D \geq \sigma_x^2 \end{cases}, \qquad \text{(A.1)}$$

the rate R_G is given in bits per source sample. The required rate is zero for a distortion equal to the variance of the source, because this distortion can be achieved by using "zero"-samples (the mean of the source-PDF) at the receiving end for reconstruction (i.e., no signal transmission is required).

Table A.1   Some probability density functions suggested in [Jayant and Noll (1984)] for investigation of rate-distortion functions and quantizer performance.

Distribution	Abbreviation	PDF $p(x)$
Uniform	U	$p(x) = \begin{cases} \frac{1}{A}, & -\frac{A}{2} < x < \frac{A}{2} \\ 0, \text{else} \end{cases}$ with $A = \sqrt{12}\,\sigma_x$
Gaussian (Normal)	G $\mathcal{N}(0, \sigma_x^2)$	$p(x) = \frac{1}{\sqrt{2\pi}\sigma_x} e^{-x^2/2\sigma_x^2}$
Laplacian	L	$p(x) = \frac{1}{\sqrt{2}\sigma_x} e^{-\sqrt{2}|x|/\sigma_x}$
Gamma	Γ	$p(x) = \frac{\sqrt[4]{3}}{\sqrt{8\pi\sigma_x|x|}} e^{-\sqrt{3}|x|/2\sigma_x}$

The DRF, the inverse of the RDF, is given by

$$D_G(R) = 2^{-2R} \cdot \sigma_x^2, \quad R \geq 0. \tag{A.2}$$

In practice, it is convenient to relate the distortion to the variance of the source. Thus, the SNR representation of the DRF is often preferred:

$$\frac{SNR_G}{\text{dB}} = 10 \log_{10} \frac{\sigma_x^2}{D_G(R)} = 10 \log_{10}(2^{2R}) \approx 6.02 \cdot R. \tag{A.3}$$

A.2.1.2   *Memoryless Non-Gaussian Sources*

The RDF $R(D)$ of any non-Gaussian source is not known analytically, but lower and upper bounds are available:

$$^L R(D) \leq R(D) \leq R_G(D). \tag{A.4}$$

The right part of the relation (A.4) says that any source with a PDF other than Gaussian requires fewer bits for coding than a Gaussian source for the same distortion D. The quantity $^L R(D)$ is called Shannon lower-bound; it will be discussed below.

For the DRF we have

$$^L D(R) \leq D(R) \leq D_G(R), \tag{A.5}$$

which, according to (A.4), means that at a given rate R the distortion D resulting for a Gaussian source is always larger than for any other source.

The Shannon lower bound for the rate is given by [Jayant and Noll (1984)]

$$^L R(D) = h(x) - \frac{1}{2} \log_2(2\pi e D) \tag{A.6}$$

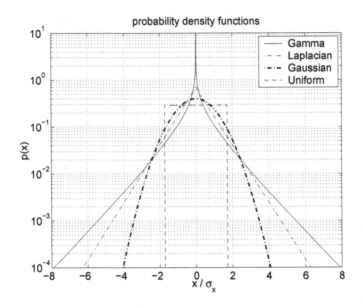

Fig. A.1   Graphs of some probability density functions.

and for the distortion it equals

$$^{L}D(R) = \frac{1}{2\pi e} 2^{-2[R-h(X)]} \ . \tag{A.7}$$

The quantity $h(X)$ is the differential entropy of the source defined by

$$h(X) \doteq \mathrm{E}_X[-\log_2 p(X)] = -\int_{-\infty}^{+\infty} p(x) \log_2 p(x) dx \ . \tag{A.8}$$

The differential entropy of any source with some given, limited variance σ_x^2 is never larger than for a Gaussian source, i.e.,

$$h(X) \leq h_G(X) = \frac{1}{2} \log_2(2\pi e \sigma_x^2) \ . \tag{A.9}$$

Thus, the Gaussian source represents the worst case for source coding from an information theoretic point-of-view. Table A.2 lists the differential entropies of several source PDFs.

The Shannon lower bound can be written in terms of SNR according to

$$\frac{SNR(^{L}D(R))}{\mathrm{dB}} = 10\log_{10}\frac{\sigma_x^2}{^{L}D(R)} = 10\log_{10}\frac{\sigma_x^2\, 2\pi e 2^{2R}}{2^{2h(X)}}$$

$$= \underbrace{R\cdot 20\log_{10}(2)}_{6.02\cdot R} + \underbrace{10\log_{10}\frac{2\pi e \sigma_x^2}{2^{2h(X)}}}_{\Delta SNR} \ , \tag{A.10}$$

Table A.2 Differential entropies of several memoryless sources [Jayant and Noll (1984)]; $C_e = 0.5772$ is the Eulerian constant.

PDF	$h(X)$	$2^{2h(X)}$
Uniform (U)	$\frac{1}{2}\log_2(12\sigma_x^2)$	$12\sigma_x^2$
Gaussian (G)	$\frac{1}{2}\log_2(2\pi e\sigma_x^2)$	$2\pi e\sigma_x^2$
Laplacian (L)	$\frac{1}{2}\log_2(2e^2\sigma_x^2)$	$2e^2\sigma_x^2$
Gamma (Γ)	$\frac{1}{2}\log_2(4\pi e^{1-C_e}\sigma_x^2/3)$	$4\pi e^{1-C_e}\sigma_x^2/3$

Table A.3 Numerical values of ΔSNR in the Shannon lower bound (A.10) for several PDFs.

U	G	L	Γ
1.533	0	0.629	4.268

i.e., the SNR for a Gaussian source is increased by some term ΔSNR, which depends on the probability distribution of the signal. Some numerical values are given in Table A.3.

Over all, we may write the DRF for any memoryless source (in terms of SNR) according to

$$6.02 \cdot R \leq SNR/\text{dB} \leq 6.02 \cdot R + \Delta SNR \qquad (A.11)$$

For high rate (e.g., $R \geq 4$ bits/sample) the DRF converges against the Shannon lower bound[Jayant and Noll (1984)], i.e., the SNR converges against the upper limit in (A.11). For low rates, accurate values of the DRF for non-Gaussian distributions are only available numerically; some examples are given in Table A.4.

A.2.2 Comparison with Practical Quantization Schemes

A.2.2.1 SNR-Values for Optimal Scalar Quantization

In Table A.5 we summarize measurement results [Jayant and Noll (1984)] for scalar optimal (Lloyd-Max) symmetric quantizers with a fixed number of bits for each source sample. The results are also valid for sources with memory as the scalar quantization is "blind" for the time-dependencies of adjacent source samples.

Table A.4   Numerical values [Jayant and Noll (1984)] of the DRF (SNR in dB) for several PDFs at low rates.

R bits/sample	U SNR/dB	G SNR/dB	L SNR/dB	Γ SNR/dB
1	6.79	6.02	6.62	8.53
2	13.21	12.04	12.66	15.21
3	19.42	18.06	18.68	21.59
$\Delta SNR|_{R\geq 4}$/dB	1.533	0	0.629	4.268

Table A.5   SNR-values in dB for optimal scalar fixed rate quantization [Jayant and Noll (1984)].

R bits/sample	U SNR/dB	G SNR/dB	L SNR/dB	Γ SNR/dB
1	6.02	4.40	3.01	1.76
2	12.04	9.30	7.54	6.35
3	18.06	14.62	12.64	11.52
4	24.08	20.22	18.13	17.07
5	30.10	26.01	23.87	22.85
6	36.12	31.89	29.74	28.73
7	42.14	37.81	35.69	34.67
$\Delta' SNR|_{R>7}$/dB	0.0	−4.347	−6.537	−7.547

For a large number of quantizer levels (high rate R), analytical results are known from asymptotic quantization theory [Gersho and Gray (2001)]:

$$SNR/\text{dB} \approx 6.02 \cdot R + \Delta' SNR , \qquad (A.12)$$

where $\Delta' SNR$ is given for our test-PDFs in Table A.5.

It is known (see Section 2.6.2.2) that the indices of optimal fixed-rate quantizers contain residual redundancies. Some numerical results for the potential gains by variable-rate index encoding (Huffman coding, [Cover and Thomas (1991)]) for optimal scalar quantizers for Gaussian samples are given in Table A.6, where $\bar{R}$ denotes the average number of bits per sample used by the Huffman code. The word-length of the fixed-rate quantizer is denoted by R in bits per sample. The SNR-values in the Table are given in dB. For comparison the SNR-values of the DRFs are also given in the rightmost column.

In general, the average rate $\bar{R}$ does not achieve the entropy $H(I)$ of the indices because a variable-rate encoded index must still have an integer

Table A.6   Potential gains by variable-rate encoding of optimal scalar quantizer indices (Gaussian source signal) [Jayant and Noll (1984)]; $H(I)$ is the entropy of the quantizer index in bits/sample.

R bits/sample	$H(I)$	$\bar{R}$ (bits/sample)	$\Delta R \rightarrow$	ΔSNR in dB	SNR in dB without ΔSNR	DRF in dB
2	1.911	1.989	0.011	0.07	9.30	12.04
3	2.825	2.884	0.116	0.70	14.62	18.06
4	3.765	3.809	0.191	1.15	20.22	24.08
5	4.730	4.771	0.229	1.38	26.01	30.10

Table A.7   Comparison of several quantization schemes with the DRF for memoryless source signals at a rate of $R = 1$ bit per sample [Jayant and Noll (1984)]. The table entries are SNR-values in dB.

PDF	VQ with vector dim. N			opt. scalar Quant. $N = 1$	DRF
	$N = 2$	$N = 4$	$N = 8$		
U	6.02*	6.04*	6.14*	6.02	6.79
G	4.5	4.7	4.8	4.40	6.02
L	3.8	4.6	4.9	3.01	6.62
Γ	3.2	5.4	6.4	1.76 (2.23)	8.53

* The SNR values for vector quantization of the uniform source are different from [Jayant and Noll (1984)], where 6.0 dB is given in each case. Since there should be some gain by VQ (as indicated by the DRF), we trained appropriate codebooks by the LBG algorithm and found the values given in the table by simulations.

number of bits, i.e., $H(I) \leq \bar{R} \leq R$. The rate-gain (and hence the SNR-gain) indicated by Table A.6 is relatively small.

A.2.2.2   Comparison of the DRF with Optimal Scalar and Vector Quantization

Table A.7 shows a comparison of several quantization schemes (without entropy coding) with the DRF for memoryless source signals at a rate of $R = 1$ bit per sample, which is a practically important case, e.g., in speech coding.

The vector quantizer codebooks were trained by the LBG algorithm [Linde *et al.* (1980)]. As this algorithm leads to a local minimum of the distortion, several training procedures with different initial codebooks were conducted and the best codebook was selected. The difference in source SNR for the different codebooks was lower than 0.1 dB.

A Gamma-PDF has many realizations that are very close to zero. Hence, if a *symmetric* scalar quantizer is used at rate of $R = 1$ bit per sample, both quantizer levels are forced to be located away from zero. Thus, the frequently occuring "close-to-zero"-realizations are quantized with large error $(SNR = 1.76 \, \mathrm{dB})$. If an asymmetric quantizer is allowed, a somewhat better performance $(SNR = 2.23 \, \mathrm{dB})$ can be achieved.

In summary, we can state that the more dissimilar a PDF is from the uniform PDF, the larger is the difference between the DRF and the practically achievable performance of a quantization scheme and, moreover, the larger is the gain by vector quantization.

A.2.3   *Gaussian Sources with Memory*

The RDF for correlated sources (and the squared-error distortion measure) is analytically only known for Gaussian PDF; therefore we will restrict the discussion to this special case. In general, the RDF and the DRF for a discrete-time correlated Gaussian source $x(k)$ are given by [Berger and Gibson (1998)]

$$D(\theta) = \frac{1}{2\pi} \int_{-\pi}^{\pi} \min\{\theta, \Phi_{xx}(\Omega)\} d\Omega \qquad (A.13)$$

$$R(\theta) = \frac{1}{2\pi} \int_{-\pi}^{\pi} \max\left\{0, \frac{1}{2} \log_2 \frac{\Phi_{xx}(\Omega)}{\theta}\right\} d\Omega \qquad (A.14)$$

with the power spectral density (PSD)

$$\Phi_{xx}(\Omega) = \mathcal{F}_*(\varphi_{xx}(\lambda)) = \sum_{\lambda=-\infty}^{+\infty} \varphi_{xx}(\lambda) \cdot e^{-j\lambda\Omega} , \qquad (A.15)$$

i.e., a particular rate-distortion pair is connected by a specific parameter θ, but there is no explicit analytical function defining, e.g., the rate for a given distortion. In (A.15), φ_{xx} denotes the autocorrelation sequence of the source signal which is defined by

$$\varphi_{xx}(\lambda) \doteq \mathrm{E}\{X(k) \cdot X(k + \lambda)\} . \qquad (A.16)$$

The formulae above are illustrated by Figure A.2 [Berger and Gibson (1998)]. While the distortion $D(\theta)$ is given by the shaded area, which corresponds to the integral in (A.13), the rate can be written as

$$R(\theta) = \frac{1}{2\pi} \int_{-\pi}^{\pi} \max\left\{0, \frac{1}{2} \log_2 \frac{\Phi_{xx}(\Omega)}{\theta}\right\} d\Omega = \frac{1}{2\pi} \int_{\Omega_A} \frac{1}{2} \log_2 \frac{\Phi_{xx}(\Omega)}{\theta} d\Omega ,$$

$$(A.17)$$

which means that only the frequency regions $\Omega \in \Omega_A$, in which the PSD of the signal is larger than the parameter θ, contribute to the rate R.

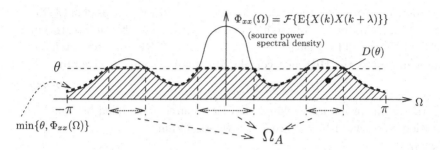

Fig. A.2 Power spectral density of a correlated signal and definition of the integration regions for the parametric computation of rate-distortion pairs for a given parameter θ.

A.2.3.1 *Simplification for High Rate*

If we assume that the distortion is low (or, equivalently, that the rate is sufficiently high), so that the PSD of the source is larger than the parameter θ at each frequency point, i.e.,

$$\theta < \Phi_{xx}(\Omega) \; \forall \, \Omega \,, \tag{A.18}$$

then the distortion (A.13) can be simplified according to

$$D(\theta) = \frac{1}{2\pi} \int_{-\pi}^{\pi} \min\{\theta, \Phi_{xx}(\Omega)\} \, d\Omega \stackrel{\theta < \Phi_{xx}}{=} \theta \tag{A.19}$$

and thus

$$R(D) = \frac{1}{2\pi} \int_{-\pi}^{\pi} \max\left\{0, \frac{1}{2} \log_2 \frac{\Phi_{xx}(\Omega)}{D}\right\} d\Omega = \frac{1}{4\pi} \int_{-\pi}^{\pi} \log_2 \frac{\Phi_{xx}(\Omega)}{D} d\Omega \,. \tag{A.20}$$

As (A.20) can be split according to

$$R(D) = \frac{1}{4\pi} \int_{-\pi}^{\pi} \log_2 \Phi_{xx}(\Omega) d\Omega \; - \; \frac{1}{2} \log_2(D) \,, \tag{A.21}$$

the DRF equals

$$D(R) = 2^{-2R} \cdot \sigma_x^2 \cdot \underbrace{\frac{2^{\int \log_2 \Phi_{xx}(\Omega)/(2\pi)}}{\sigma_x^2}}_{\doteq \gamma_x^2} \,. \tag{A.22}$$

In (A.22) the spectral flatness measure (SFM)

$$\gamma_x^2 \doteq \exp\left(\frac{1}{2\pi} \int_{-\pi}^{+\pi} \log_e \Phi_{xx}(\Omega) d\Omega\right) \Big/ \underbrace{\frac{1}{2\pi} \int_{-\pi}^{+\pi} \Phi_{xx}(\Omega) d\Omega}_{\sigma_x^2} \,, \quad 0 \le \gamma_x^2 \le 1 \tag{A.23}$$

is introduced. Note that the natural logarithm is equivalently used in (A.23) and that the variance σ_x^2 of the input signal is expressed by an integration of the PSD over all frequencies. As $2^{-2R} \cdot \sigma_x^2$ is the DRF of an *uncorrelated* Gaussian source, the DRF of a correlated Gaussian source is given by

$$D(R) = D(R)|_{\text{uncorrelated}} \cdot \gamma_x^2 , \qquad (A.24)$$

i.e., the result for an uncorrelated Gaussian source is modified by a multiplication with the SFM. In terms of SNR we obtain

$$\frac{SNR}{\text{dB}} = 6.02 \cdot R + \underbrace{10 \log_{10}\left(\frac{1}{\gamma_x^2}\right)}_{\text{additive term, due to correlation}} \qquad \text{with} \quad 0 \le \gamma_x^2 \le 1 \quad (A.25)$$

For an uncorrelated signal the spectrum $\Phi_{xx}(\Omega)$ is totally flat, i.e., $\gamma_x^2 = 1$ and there is no SNR gain. If γ_x^2 is close to zero, the source signal is strongly correlated and a large SNR-gain is introduced in (A.25) by the additive term.

A.2.3.2 *Simplification for High Rate and a Linearly Filtered Gaussian Source*

In what follows we will assume that the correlated source is generated by filtering a Gaussian source according to Figure A.3. The filter is assumed to

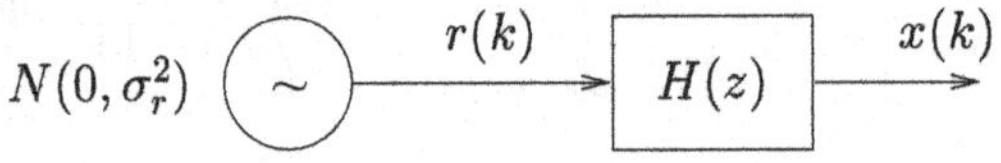

Fig. A.3   Gauss-Markov source.

be linear, stable, and of minimum-phase [Schuessler (1991)]. The transfer function of such a linear filter can be written as

$$H(z) = \left(1 + \sum_{i=1}^{N_z} b_i \cdot z^{-i}\right) \bigg/ \left(1 + \sum_{i=1}^{N_p} a_i \cdot z^{-i}\right) , \qquad N_z \le N_p . \quad (A.26)$$

The excitation signal $r(k)$ in Figure A.3 is assumed to be stationary, zero mean, uncorrelated, and normally distributed. Then, the autocorrelation sequence equals

$$\varphi_{rr}(\lambda) = E\{R(k) \cdot R(k+\lambda)\} = \sigma_r^2 \cdot \begin{cases} 1, \lambda = 0 \\ 0, \text{else} \end{cases} \qquad (A.27)$$

and its Fourier-transform, the power spectral density, is

$$\Phi_{rr}(\Omega) = F_*(\varphi_{rr}(\lambda)) = \sigma_r^2 \, , \quad -\pi \le \Omega \le +\pi \, . \tag{A.28}$$

At the output of the filter we obtain the PSD

$$\Phi_{xx}(\Omega) = |H(\Omega)|^2 \cdot \Phi_{rr}(\Omega) = |H(\Omega)|^2 \cdot \sigma_r^2 \, , \tag{A.29}$$

and the signal variance at the filter output is given by

$$\sigma_x^2 = \frac{1}{2\pi} \int\limits_{-\pi}^{+\pi} \Phi_{xx}(\Omega) d\Omega = \sigma_r^2 \cdot \frac{1}{2\pi} \int\limits_{-\pi}^{+\pi} |H(\Omega)|^2 \, d\Omega \, . \tag{A.30}$$

By use of (A.23) we obtain

$$\gamma_x^2 = \frac{\exp\left(\dfrac{1}{2\pi} \int\limits_{-\pi}^{+\pi} \log_e\left(\sigma_r^2 \cdot |H(\Omega)|^2\right) d\Omega\right)}{\sigma_x^2}$$

$$= \frac{\sigma_r^2 \cdot \exp\left(\dfrac{1}{2\pi} \int\limits_{-\pi}^{+\pi} \log_e\left(|H(\Omega)|^2\right) d\Omega\right)}{\sigma_x^2} \tag{A.31}$$

One can show [Jayant and Noll (1984)] that

$$\frac{1}{2\pi} \int\limits_{-\pi}^{+\pi} \log_e\left(|H(\Omega)|^2\right) d\Omega = 0 \tag{A.32}$$

if the poles and zeros of the transfer function $H(z)$ of the linear filter are located within the unit-circle, i.e., the filter must be stable and have minimum phase (as assumed above).

Using (A.30) and (A.32) we obtain from (A.31)

$$\frac{1}{\gamma_x^2} = \frac{\sigma_x^2}{\sigma_r^2} = \frac{1}{2\pi} \int\limits_{-\pi}^{+\pi} |H(\Omega)|^2 \, d\Omega = \sum_{k=0}^{\infty} h_0^2(k) \, , \tag{A.33}$$

where $h_0(k)$ is the impulse response of the filter, i.e., $h_0(k) = Z^{-1}\{H(z)\}$; for the rightmost equality Parseval's theorem was used. The condition $h_0^2(0) = 1$, which is required for (A.32) to hold, is always fulfilled by (A.26).

Example: Let us assume a discrete-time uncorrelated Gaussian source that is filtered by a filter with the transfer function

$$H(z) = \frac{z}{z-a} \quad \text{with} \quad a = 0.9 \,. \tag{A.34}$$

The corresponding impulse response equals

$$h_0(k) = Z^{-1}\{H(z)\} = \begin{cases} a^k, \ k \geq 0 \\ 0, \quad \text{else} \end{cases} . \tag{A.35}$$

Hence,

$$\frac{1}{\gamma_x^2} = \sum_{k=0}^{\infty} h_0^2(k) = \sum_{k=0}^{\infty} a^{2k} = \frac{1}{1-a^2} \tag{A.36}$$

and we obtain for the spectral flatness measure:

$$\gamma_x^2 = 1 - a^2 \,. \tag{A.37}$$

Thus, the SNR-version of the DRF equals

$$\frac{SNR}{\text{dB}} = 6.02 \cdot R + 10 \log_{10}\left(\frac{1}{\gamma_x^2}\right) \overset{a=0.9}{=} 6.02 \cdot R + \underbrace{7.21}_{\text{gain}}, \tag{A.38}$$

i.e., we have a gain of 7.21 dB due to the correlation of the source.

A.2.3.3 *Quality of the High-Rate Approximation*

The high-rate approximation (A.38) is an *upper* bound of the true SNR performance limit, because in (A.18) we have assumed that the parameter θ is lower than the PSD of the signal at each frequency, which is not true for low rates. Therefore, we neglect parts of the distortion and, thus, the SNR-curve over-estimates the true performance for low rates (this can be easily seen if we insert $R = 0$ into (A.38): the result should be 0 dB). The over-estimation of the performance is illustrated by Figure A.4, where the exact DRFs (SNR-versions) and the high-rate approximations are compared for examples of first and second order Gauss-Markov sources.

The transfer functions $H(z)$ of the filters used to generate the correlated source-signal are also given in the plot. The power spectral densities $\Phi_{xx}(\Omega)$ of the signals $x(k)$ resulting from filtering are plotted in Figure A.5. They directly result from the frequency responses of the filters according to (A.29).

The Figures A.4 and A.5 indicate that, even if the PSD of the signal has sharp peaks, the high-rate approximation of the DRF works well, as long as the rate R is above one bit per sample. The approximation is bad, however, for low rates such as $R = 0.25$ bits per sample. Moreover, Figure A.4 shows that if a strongly correlated source shall be coded with very low bit rate (e.g., $R < 0.2$ bits per sample) it is *principally* impossible to exploit the full auto-correlation for quality improvement.

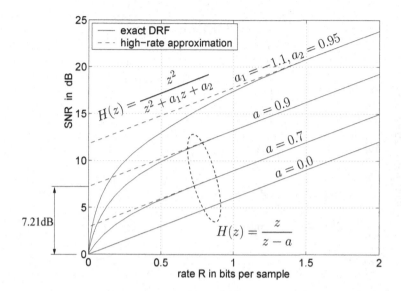

Fig. A.4   DRFs (SNR versions) of auto-correlated Gaussian sources.

A.3   Capacities of Practically Important Channels

The channel capacity quantifies the number of bits per channel use (or per time unit) that can on average be transmitted over a (noisy) channel without errors. From an information-theoretic point-of-view, not the reliability but the amount (measured in bits) of transmitted information is limited by the channel noise.

In what follows we summarize the capacities for a number of practically important channels.

A.3.1   *Binary Symmetric Channel (BSC)*

The simplest channel model is the binary symmetric channel (BSC), which is depicted in Figure A.6. The input bits $z \in \{0, 1\}$ are flipped with the probability p and at the output we receive the bits $\tilde{z} \in \{0, 1\}$. The channel is assumed to be memoryless, i.e., each individual bit is transmitted independently of earlier or future error-events on the channel. The error events are also independent of the data bits.

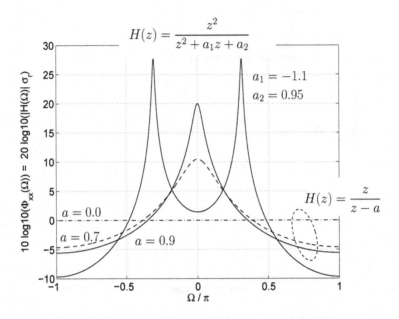

Fig. A.5 Power spectral densities of filtered Gaussian sources, with $\sigma_r^2 = 1$ the variance of the Gaussian excitation signal.

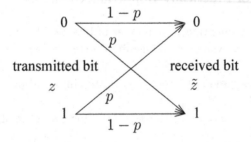

Fig. A.6 Binary symmetric channel (BSC).

The capacity of this channel equals

$$C = 1 - H(p) \quad \text{in} \quad \frac{\text{bits}}{\text{channel-use}} \tag{A.39}$$

with the binary entropy function

$$H(p) = -p \log_2(p) - (1 - p) \log_2(1 - p) . \tag{A.40}$$

In Figure A.7 the capacity of the BSC is shown versus the bit error probability p. The capacity (A.39) is achieved with independent and uniformly

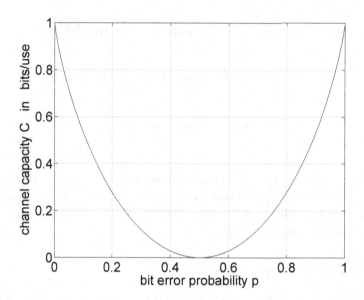

Fig. A.7 Channel capacity of a binary symmetric channel.

distributed input bits.

A.3.2 *Binary Erasure Channel (BEC)*

On a binary erasure channel, which is depicted in Figure A.8, the transmitted bit z is erased with some probability p, and when this happens, it is known at the receiver.

The capacity of this channel equals

$$C = 1 - p \quad \text{in} \quad \frac{\text{bits}}{\text{channel-use}} \tag{A.41}$$

and it is achieved with independent and uniformly distributed input bits.

A.3.3 *Additive White Gaussian Noise (AWGN) Channel*

If information is transmitted over an AWGN channel, independent Gaussian noise is added to the transmitted channel symbols. As the noise is assumed to be "white," the channel is memoryless. Several subtypes of this

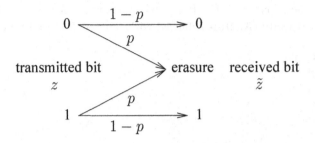

Fig. A.8   Binary erasure channel (BEC).

channel exist, depending on the channel input alphabet that is selected. All practically useful channel models with continuous inputs have in common that their input power is limited.[2] For brevity we omit this "power constraint" in the headings below.

A.3.3.1   *Discrete-Time AWGN Channel*

Real data symbols z_k (k is the discrete time index) have to be transmitted with a maximum average power P (i.e., $E\{Z_k^2\} \leq P$) over a channel that adds independent Gaussian noise samples w_k with zero mean and the variance σ_w^2. This channel is depicted in Figure A.9. Its channel capacity is

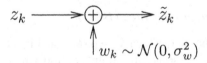

Fig. A.9   Discrete-time AWGN channel.

given by [Cover and Thomas (1991)]

$$C = \frac{1}{2}\log_2\left(1 + \frac{P}{\sigma_w^2}\right) \quad \text{in} \quad \frac{\text{bits}}{\text{channel-use}}, \quad (A.42)$$

and it is achieved by a Gaussian input distribution, i.e., $Z_k \sim \mathcal{N}(0, P)$.

---

[2]Without an input power constraint, we could achieve error-free transmission by use of infinitely large power.

A.3.3.2 *Continuous-Time Band-Limited AWGN Channel*

The data symbols have to be transmitted by continuous-time waveforms because we have a continuous-time, real-input transmission channel. The bandwidth B (positive frequencies) we are allowed to use is limited as well as the average power at the input, i.e., $E\{Z^2\} \leq P$. On the channel, continuous-time additive white Gaussian noise corrupts the transmitted information; the noise has the power spectral density $\Phi_{ww}(f) = N_0/2$ for all (positive and negative) frequencies.

As we require the channel-input signal $z(t)$ to be band-limited, it can be written as

$$z(t) = \sum_{k=-\infty}^{+\infty} z(kT_s) \cdot \mathrm{si}\left(2\pi B(t - kT_s)\right) \quad \text{with} \quad T_s = \frac{1}{2B}, \qquad (A.43)$$

i.e., samples $z_k \doteq z(kT_s)$, which are taken from the continuous-time signal with the period $T_s = \frac{1}{2B}$, are enough to completely define the continuous-time signal $z(t)$ with bandwidth B (sampling theorem). In (A.43) si-functions ($\mathrm{si}(x) \doteq \sin(x)/x$) are used to interpolate between the given samples z_k of the continuous-time signal to get the value $z(t)$ for any t. In the context of information transmission the sampling theorem means that we may select $2B$/Hz real data samples z_k per second as the information to be transmitted, and this is the maximum rate at which we can transmit.

Figure A.10 shows a transmission system in which we use $z(t)$ according to (A.43) as the channel input signal. The si-function is used as the impulse-

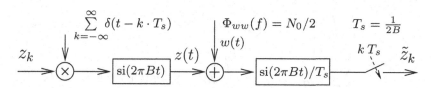

Fig. A.10 Information transmission over a continuous-time band-limited channel.

response of a pulse-shaping filter that generates the band-limited channel input signal $z(t)$ from the samples z_k, which form the actual "user data." As we have to fulfill the power constraint, we need to know the relation between the samples z_k and the power of the channel input signal $z(t)$.

Average Transmission Power: We will use the auto-correlation function (ACF)

$$\varphi_{zz}(t, t+\tau) \doteq \mathrm{E}\{Z(t)Z(t+\tau)\} \tag{A.44}$$

to compute the average power of the continuous-time signal $z(t)$. If we insert (A.43) into (A.44) we obtain

$$\varphi_{zz}(t, t+\tau) = \sum_{k=-\infty}^{\infty} \sum_{j=-\infty}^{\infty} \mathrm{E}\{Z_k Z_j\} g(t - kT_s) g(t + \tau - jT_s) , \tag{A.45}$$

where, for brevity, we use $g(t) \doteq \mathrm{si}(2\pi Bt)$. We assume that the random process of the discrete input samples z_k is at least wide-sense stationary [Proakis (1995)], i.e.,

$$\phi_{zz}(k, j) \doteq \mathrm{E}\{Z_k Z_j\} = \phi_{zz}(j - k) , \tag{A.46}$$

which means that the auto-correlation sequence $\phi_{zz}(k, j)$ only depends on the time span between its two arguments. If we use that in (A.45) and substitute $j' = j - k$, we obtain

$$\varphi_{zz}(t, t+\tau) = \sum_{j'=-\infty}^{\infty} \phi_{zz}(j') \underbrace{\sum_{k=-\infty}^{\infty} g(t - kT_s) g(t + \tau - j'T_s - kT_s)}_{c(t)} .$$

$$\tag{A.47}$$

The function $c(t)$ is periodic in t with the period T_s, i.e., it is cyclo-stationary [Proakis (1995)]. Hence, the ACF $\varphi_{zz}(t, t+\tau)$ is also periodic in t and it does not depend only on τ. As we are interested in the *average* ACF (in order to compute the average power at the channel input) the ACF (A.47) is averaged over one symbol period T_s:

$$\overline{\varphi}_{zz}(\tau) \quad \doteq \quad \frac{1}{T_s} \int_{T_s/2}^{-T_s/2} \varphi_{zz}(t, t+\tau) dt \tag{A.48}$$

$$\stackrel{(A.47)}{=} \sum_{j'=-\infty}^{\infty} \frac{\phi_{zz}(j')}{T_s} \sum_{k=-\infty}^{\infty} \int_{-T_s/2}^{T_s/2} g(t - kT_s) g(t + \tau - j'T_s - kT_s) dt$$

$$\stackrel{t'=t-kT_s}{=} \sum_{j'=-\infty}^{\infty} \frac{\phi_{zz}(j')}{T_s} \sum_{k=-\infty}^{\infty} \int_{-T_s/2-kT_s}^{T_s/2-kT_s} g(t') g(t' + \tau - j'T_s) dt'$$

$$= \sum_{j'=-\infty}^{\infty} \phi_{zz}(j') \frac{1}{T_s} \int_{-\infty}^{+\infty} g(t) g(t + \tau - j'T_s) dt . \tag{A.49}$$

We obtain the desired average power of the continuous-time signal $z(t)$ for $\tau = 0$:

$$P = \overline{\varphi}_{zz}(0) = \sum_{j=-\infty}^{\infty} \phi_{zz}(j) \frac{1}{T_s} \int_{-\infty}^{+\infty} g(t)g(t - jT_s)dt \ . \qquad (A.50)$$

If, in place of $g(t)$, we re-insert the si-functions back into (A.50) and consider the orthogonality relation

$$\frac{1}{T_s} \int_{-\infty}^{+\infty} \text{si}\left(2\pi B(\tau - jT_s)\right) \cdot \text{si}\left(2\pi B(\tau - kT_s)\right) d\tau = \begin{cases} 1 \ k = j \\ 0 \ k \neq i \end{cases} , \qquad (A.51)$$

we obtain

$$P = \overline{\varphi}_{zz}(0) = \phi_{zz}(0) = \text{E}\{Z_k^2\} \ . \qquad (A.52)$$

This means that in our system in Figure A.10 we can compute the average transmission power of the continuous-time signal $z(t)$ directly from the power of the discrete-time data samples z_k, and we can easily apply the power constraint directly to the data samples.

Receiver: If there was no noise on the channel, the easiest way to reconstruct the data samples at the receiver would be to re-sample the signal $z(t)$ at the times kT_s. But there is AWGN on the channel and it is known (e.g., [Haykin (2000)], [Proakis (1995)]) that a filter matched to the transmission pulse $g(t)$ (the si-function in our case) must be used at the receiver to achieve the optimum signal-to-noise ratio at sampling time. In our case, the matched filter has the impulse response

$$h(t) = \frac{1}{T_s}g^*(-t) = \frac{1}{T_s}\text{si}\left(-2\pi Bt\right) = \frac{1}{T_s}\text{si}\left(2\pi Bt\right) \ . \qquad (A.53)$$

Hence, the sampled matched filter output equals

$$\tilde{z}_k \quad = \quad \left. (z(t) + w(t)) * \frac{1}{T_s}\text{si}\left(2\pi Bt\right) \right|_{t=kT_s} \qquad (A.54)$$

$$= \quad \frac{1}{T_s} \int_{-\infty}^{+\infty} (z(\tau) + w(\tau)) \cdot \text{si}\left(2\pi B(kT_s - \tau)\right) d\tau \qquad (A.55)$$

$$\overset{(A.43)}{=} \sum_{j=-\infty}^{+\infty} z(jT_s) \cdot \frac{1}{T_s} \int_{-\infty}^{+\infty} \text{si}\left(2\pi B(\tau - jT_s)\right) \cdot \text{si}\left(2\pi B(kT_s - \tau)\right) d\tau$$

$$+ \frac{1}{T_s} \int_{-\infty}^{+\infty} w(\tau) \cdot \text{si}\left(2\pi B(kT_s - \tau)\right) d\tau \ . \qquad (A.56)$$

Since the si-function is even and by virtue of (A.51) we obtain

$$\tilde{z}_k = z_k + w_k \ , \tag{A.57}$$

with

$$w_k = \frac{1}{T_s} \int\limits_{-\infty}^{+\infty} w(\tau) \cdot \text{si} \left(2\pi B(\tau - kT_s) \right) d\tau \ , \tag{A.58}$$

which means that we have found an equivalent discrete-time channel model.

The (linearly) filtered noise samples w_k are known to be independent and Gaussian distributed with zero mean [Proakis (1995)], as the filter input process $w(t)$ is Gaussian and zero mean with the PSD $\Phi_{ww}(f) = N_0/2$, from which we obtain the ACF

$$\varphi_{ww}(\tau) \doteq \text{E}\{W(t)W(t+\tau)\} = \mathcal{F}^{-1}(\Phi_{ww}(f)) = \frac{N_0}{2} \delta(\tau) \ . \tag{A.59}$$

The variance $\sigma_w^2 = \text{E}\{W_k^2\}$ of the noise samples w_k is given by

$$\sigma_w^2 = \frac{1}{T_s^2} \int\limits_{-\infty}^{+\infty} \int\limits_{-\infty}^{+\infty} \text{E}\{W(\tau)W(\tau')\} \cdot \text{si} \left(2\pi B(\tau - kT_s) \right) \text{si} \left(2\pi B(\tau' - kT_s) \right) d\tau d\tau'. \tag{A.60}$$

If we substitute $\tau' = \tau + \xi$ and insert the ACF (A.59) of the noise we obtain

$$
\begin{aligned}
\sigma_w^2 &= \frac{1}{T_s^2} \int\limits_{-\infty}^{+\infty} \int\limits_{-\infty}^{+\infty} \text{E}\{W(\tau) \cdot W(\tau + \xi)\} \cdot \\
&\qquad\qquad \cdot \text{si} \left(2\pi B(\tau - kT_s) \right) \text{si} \left(2\pi B(\tau + \xi - kT_s) \right) d\tau d\xi \\
&= \frac{N_0}{2T_s^2} \int\limits_{-\infty}^{+\infty} \text{si} \left(2\pi B(\tau - kT_s) \right) \left(\int\limits_{-\infty}^{+\infty} \delta(\xi) \cdot \text{si} \left(2\pi B(\tau + \xi - kT_s) \right) d\xi \right) d\tau \\
&= \frac{N_0}{2T_s^2} \int\limits_{-\infty}^{+\infty} \text{si}^2 (2\pi B(\tau - kT_s)) d\tau = \frac{N_0}{2T_s^2} \cdot \frac{1}{2B} \\
&= \frac{N_0}{2T_s} = N_0 B \ . \tag{A.61}
\end{aligned}
$$

Hence, we constructed band-limited signals for the transmission over the continuous-time AWGN channel and we obtained an equivalent discrete-time channel model, in which the channel input power equals $P = \text{E}\{Z_k^2\}$ and the variance of the discrete-time additive white Gaussian noise samples equals $\sigma_w^2 = \frac{N_0}{2T_s}$, with the symbol transmission period $T_s = \frac{1}{2B}$ and the system bandwidth B. Therefore, we can use (A.42) to find the channel capacity of the band-limited continuous-time AWGN channel:

$$C = \frac{1}{2} \log_2 \left(1 + \frac{2T_s P}{N_0} \right) \quad \text{in} \quad \frac{\text{bits}}{\text{channel-use}} . \qquad (A.62)$$

If we introduce the symbol energy

$$E_s = PT_s \qquad (A.63)$$

that, on average, is used for the transmission of each data sample z_k, we find

$$C = \frac{1}{2} \log_2 \left(1 + \frac{2E_s}{N_0} \right) \quad \text{in} \quad \frac{\text{bits}}{\text{channel-use}} . \qquad (A.64)$$

This capacity is achieved by a Gaussian input distribution, i.e., $Z_k \sim \mathcal{N}(0, P)$.

The capacity (A.64) is often related to the bandwidth-constraint B. As the bandwidth determines the symbol transmission period $T_s = \frac{1}{2B}$ (the time between two channel uses) we can also write

$$C' = \frac{1}{T_s} \cdot \frac{1}{2} \log_2 \left(1 + \frac{2E_s}{N_0} \right) = B \log_2 \left(1 + \frac{P}{N_0 B} \right) \quad \text{in} \quad \frac{\text{bits}}{\text{s}} . \qquad (A.65)$$

Note that the capacity C' increases only logarithmically with the transmission power. If, however, we use a very large bandwidth we obtain

$$C'_\infty = \lim_{B \to \infty} C' = \lim_{B \to \infty} B \log_2 \left(1 + \frac{P}{N_0 B} \right) = \frac{P}{N_0 \log_e(2)} \quad \text{in} \quad \frac{\text{bits}}{\text{s}} , \qquad (A.66)$$

(L'Hospital's rule) i.e., the channel capacity grows linearly with the power.

A.3.3.3 *Discrete-Time Two-Dimensional AWGN Channel*

This vector channel has two independent discrete-time real AWGN channel components, over which we transmit two-dimensional data symbols $z_k = \{z_{I,k}, z_{Q,k}\}$. Sometimes the two real data-symbol components are, for convenience of notation, arranged in a complex number as its real and imaginary parts and, consequently, the channel is then assumed to be adding "complex" noise, hence the terms "complex channel" and "complex modulation". In this book, however, we will prefer the vector notation.

Furthermore, the sum of the average transmission powers used on both real channel components is assumed to be limited, i.e., $\mathrm{E}\{\|Z_k\|^2\} = \mathrm{E}\{Z_{I,k}^2\} + \mathrm{E}\{Z_{Q,k}^2\} \le P$. On both real channel components, zero mean independent Gaussian noise samples $w_{I,k}, w_{Q,k}$, both with the variance σ_w^2, are added to the data symbols.

Due to the symmetry of the problem, the optimal way to distribute the total power P on the vector channel is to use half of it for the transmission over each of the two real channel components. Thus, the capacity of the vector channel can be computed by use of (A.42) according to

$$C = 2 \cdot \frac{1}{2} \log_2 \left(1 + \frac{P/2}{\sigma_w^2} \right) = \log_2 \left(1 + \frac{P}{2\sigma_w^2} \right) \qquad \text{in} \qquad \frac{\text{bits}}{\text{channel-use}} \tag{A.67}$$

and it is achieved by independent Gaussian input symbols in both channels according to $Z_{I,k} \sim \mathcal{N}(0, P/2)$ and $Z_{Q,k} \sim \mathcal{N}(0, P/2)$.

Such a vector channel is practically important, because the modulation of band-limited user data (bandwidth B) onto the orthogonal sinusoidal carriers $\sin(2\pi f_0 t)$ and $\cos(2\pi f_0 t)$—both with a carrier frequency $f_0 \gg B$—offers the opportunity to use two independent signal channels for transmission. The data transmitted via the cosine-carrier is often referred to as "in-phase component" and the data transmitted via the sine-carrier is denoted by "quadrature component." For convenience, both components are often interpreted as components of complex numbers, i.e., $z_k = z_{I,k} + j z_{Q,k}$, and the high-frequency transmission systems are treated as equivalent complex baseband systems with complex input and output signals. In this work, however, we use the real vector notation.

Similar as in the previous section, the power and the variance are often expressed in terms of $E_s = P \cdot T_s$ and N_0 of an equivalent continuous-time band-limited channel (positive frequencies limited to B in each component). Hence, if we assume that the noise power spectral density equals $\Phi_{ww}(f) = N_0/2$ (i.e., $\sigma_w^2 = \frac{N_0}{2T_s}$ as in (A.61)) on both continuous-time channel components, we obtain from (A.67)

$$C = \log_2 \left(1 + \frac{E_s}{N_0} \right) \qquad \text{in} \qquad \frac{\text{bits}}{\text{channel-use}} . \tag{A.68}$$

A.3.3.4   *Equivalent Power-Normalized Channel Model for Discrete Input Alphabets*

In practice, the channel inputs z_k are often taken from a limited number M of alternatives, i.e., $z_k \in \mathcal{S}_M$, where the set $\mathcal{S}_M \doteq \{s_0, s_1, ..., s_{M-1}\}$ is usually called "modulation signal set" with the "modulation signal points" s_i. Groups of $\log_2(M)$ input bits—the actual data to be transmitted—are used to address one point out of the signal set, and this signal point is transmitted over the channel.

At the receiver the goal is to correctly detect the transmitted signal point, in order to output its bit-labels as the user data. For this, the a-posteriori probability (APP) of each signal point must be computed; a hard detection would be given by a decision in favor of the most probable point. If advanced receiver concepts are used, hard decisions are *not* taken because the reliability information given by the APPs can be exploited by successive processing steps (e.g., channel decoding) with large performance gains [Hagenauer (1992)].

Figure A.11 shows the system concept mentioned above; a so-called 8-PSK (phase shift keying) modulation signal set is used as an example. At

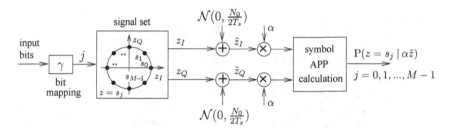

Fig. A.11   Transmission system for discrete channel inputs.

the receiver we have introduced the factor $\alpha > 0$. In what follows we will investigate the influence of the factor on the output APPs, which we can compute as follows:

$$\mathrm{P}(z = s_j \mid \alpha\tilde{z}) = \frac{p(z = s_j, \alpha \cdot \tilde{z})}{p(\alpha\tilde{z})} = \frac{p(\alpha\tilde{z} \mid z = s_j)\mathrm{P}(z = s_j)}{\sum\limits_{j'=0}^{M-1} p(\alpha\tilde{z} \mid z = s'_j)\mathrm{P}(z = s'_j)} . \quad (A.69)$$

As we use a two-dimensional vector channel with independent components, the "channel PDF" can be separated according to

$$p(\alpha\tilde{z} \mid z = s_j) = p(\alpha\tilde{z}_I \mid z_I = s_{I,j}) \cdot p(\alpha\tilde{z}_Q \mid z_Q = s_{Q,j}) . \quad (A.70)$$

Without the factor, the PDFs of the channel components are given by Gaussian distributions, e.g.,

$$p_c(\tilde{z}_I \mid z_I) = \frac{1}{\sqrt{2\pi}\sigma_w} e^{-(\tilde{z}_I - z_I)^2/(2\sigma_w^2)} \quad (A.71)$$

with $\sigma_w^2 = \frac{N_0}{2T_s}$. The PDF $p_Y(y)$ of a random variable Y that is a monotone function $y = g(x)$ of another random variable X with known PDF $p_X(x)$

is known to equal

$$p_Y(y) = p_X(g^{-1}(y)) \cdot \left| \frac{d(g^{-1}(y))}{dy} \right|, \qquad (A.72)$$

where $g^{-1}(y)$ is the function inverse to $y = g(x)$. In our case, $y = \alpha \tilde{z}$, i.e.,

$$p(\alpha \tilde{z}_I \mid z_I) = \frac{1}{\alpha} \, p_c(\tilde{z}_I \mid z_I) . \qquad (A.73)$$

If we insert (A.73) into (A.70) and the result into (A.69), we see that the factor α cancels out and $\mathrm{P}(z = s_j \mid \alpha \tilde{z}) = \mathrm{P}(z = s_j \mid \tilde{z})$. Hence, we are free to choose the factor $\alpha > 0$.

Very often, $\alpha = \frac{1}{\sqrt{P}}$ is selected and the factor is shifted backwards over the noise summation points in Figure A.11, which leads to the system depicted in Figure A.12. As $E_s = P \cdot T_s$, the variance of the additive noise is

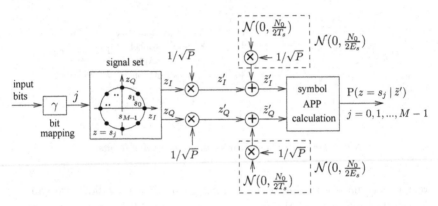

Fig. A.12   Transmission system with normalized noise power.

scaled to $\sigma_n^2 = \frac{N_0}{2E_s}$, i.e., the "channel-quality" parameter E_s/N_0 is directly contained in the noise variance, which is used in both channel components for a simulation.

The input power $\mathrm{E}\{\|Z'\|^2\}$ of the equivalent channel is normalized to one if $\alpha = 1/\sqrt{P}$, i.e., the power in each real channel dimension equals $\mathrm{E}\{Z_I'^2\} = \mathrm{E}\{Z_Q'^2\} = 1/2$; the scaling factor can be integrated into the signal constellation, so that, in a simulation, the latter must fulfill the condition

$$\mathrm{E}\{\|Z'\|^2\} = \sum_{j=0}^{M-1} \left(s_{I,j}^2 + s_{Q,j}^2 \right) \mathrm{P}(z' = s_j) \overset{!}{=} 1 . \qquad (A.74)$$

In binary data transmission, the probability distribution of the signal points is often uniform, i.e., $\mathrm{P}(z' = s_j) = 1/M$, as the input bits are independent and uniformly distributed.

In what follows, we give the channel capacities for a number of AWGN channels with discrete input alphabets. In all cases we assume that the equivalent normalized channel described in this section is used.

A.3.3.5 *Discrete-Time AWGN Channel with Binary Input*

We assume that the input bits are transmitted by the two signal points $z_k = \pm 1$. Then, the channel capacity equals

$$C = \frac{1}{2} \sum_{z \in \{-1,+1\}} \int_{-\infty}^{\infty} p_N(\tilde{z}|z) \log_2 \frac{2 \cdot p_N(\tilde{z}|z)}{p_N(\tilde{z}|z=-1) + p_N(\tilde{z}|z=1)} d\tilde{z} \quad (A.75)$$

with

$$p_N(\tilde{z}|z) = \frac{1}{\sqrt{2\pi}\sigma_n} e^{-\frac{1}{2\sigma_n^2}(\tilde{z}-z)^2} \quad \text{with} \quad \sigma_n^2 = \frac{N_0}{2E_s}. \quad (A.76)$$

Due to the symmetry of the problem, the capacity *is* achieved with equally probable input bits, which is implicitly assumed in (A.75). As stated in the previous section, the channel input power is normalized to one, i.e., $E\{Z^2\} = 1$, and the variance of the noise is scaled with the inverse of the power constraint. Unfortunately, (A.75) can not be solved analytically, i.e., for a given value of E_s/N_0 we have to find the capacity numerically.

If we use a binary-input AWGN channel with hard decisions at the output, this channel can be modeled by a BSC with a bit error probability of $p = \frac{1}{2}\text{erfc}\sqrt{E_s/N_0}$ [Proakis (1995); Haykin (2000)]. We compare the capacities of binary-input, discrete-time AWGN channels with continuous and with binary outputs in Figure A.13; we observe a maximum gain of about 2 dB if we do *not* hard-decide the symbols at the channel output.

A.3.3.6 *Discrete-Time AWGN Channel with Discrete Input*

We assume M input symbols from the modulation signal set $\mathcal{S}_M$ and, as we have a real channel (one component), this is amplitude shift keying modulation (M-ASK). Usually, a uniform probability distribution of the signal points is additionally required and this leads to the problem that we cannot achieve the true capacity of this channel, since we know that on a one-dimensional AWGN-channel the inputs must have a Gaussian distribution. As the data bits we want to transmit are independent and uniformly distributed, we accept this as another system constraint and we compute the "constrained" capacity according to

$$C = \frac{1}{M} \sum_{z \in \mathcal{S}_M} \int_{-\infty}^{\infty} p_N(\tilde{z}|z) \log_2 \frac{M \cdot p_N(\tilde{z}|z)}{\sum_{z' \in \mathcal{S}_M} p_N(\tilde{z}|z')} d\tilde{z} \quad (A.77)$$

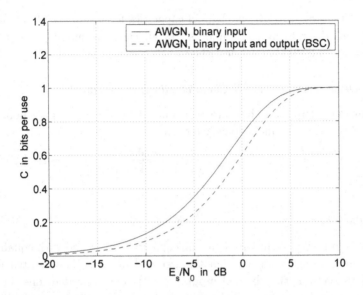

Fig. A.13   Capacities of binary-input discrete-time AWGN channels with continuous and binary outputs. The latter case is equivalent to a binary symmetric channel (BSC) with $p_e = \frac{1}{2}\mathrm{erfc}\sqrt{E_s/N_0}$.

with $p_N(\tilde{z}|z)$ given by (A.76). It should be noticed that (A.77) is nothing but the result of the direct evaluation of the definition of the mutual information [Cover and Thomas (1991)] for a given input probability distribution, i.e., the maximization of the mutual information over the input probability distribution, which is part of the definition of the channel capacity, has been omitted. The constrained capacity given by (A.77) describes the amount of information that can be transmitted if we do have a uniform probability distribution at the channel input. Hence, (A.77) is exactly what we want know.

A.3.3.7   *Discrete Time Two-Dimensional AWGN Channel with Discrete Inputs*

This channel is a two-dimensional extension of the one in the previous section. The M two-dimensional modulation signal points are selected from the set $\mathcal{S}_M$, and we again assume a uniform probability distribution of the signal points. Similar as above, the constrained capacity is given by

$$C = \frac{1}{M} \sum_{\{z_I, z_Q\} \in \mathcal{S}_M} \int_{-\infty}^{\infty} \int_{-\infty}^{\infty} p_N(\tilde{z}_I | z_I) \cdot p_N(\tilde{z}_Q | z_Q) \cdot$$

$$\cdot \log_2 \frac{M \cdot p_N(\tilde{z}_I | z_I) \cdot p_N(\tilde{z}_Q | z_Q)}{\sum_{\{z_I', z_Q'\} \in \mathcal{S}_M} p_N(\tilde{z}_I | z_I') \cdot p_N(\tilde{z}_Q | z_Q')} d\tilde{z}_I d\tilde{z}_Q , \quad \text{(A.78)}$$

with $p_N(\cdot|\cdot)$ given by (A.76), i.e., the Gaussian noise added by each channel component is assumed to be white and mutually independent with the variance $\sigma_n^2 = \frac{N_0}{2E_s}$.

In general, the modulation signal points are located "somewhere" in the two-dimensional space and then we call the transmission scheme quadrature amplitude modulation (QAM). For this case, (A.78) again does not give the true channel capacity as we assume a uniform symbol distribution. The latter assumption *is* capacity achieving if the modulation signal points all have the same power, i.e., if they are all located—with equal distances— on a circle of radius one which is centered in the origin: this transmission scheme is usually called M-ary phase shift keying (M-PSK).

A.3.3.8 *Some Capacity-Curves for M-PSK*

In Figure A.14 we show some capacity curves for AWGN channels with M-PSK inputs, which we found by numerical evaluation of (A.78). For comparison, we also included the capacity curves for one- and two-dimensional AWGN channels with unconstrained inputs, which are given by (A.64) and (A.68) respectively. Figure A.14 makes clear that the use of more than 8 PSK signal points does not provide larger channel capacity for E_s/N_0 lower than 10 dB. It should be noticed, however, that this is only true from an information-theoretic point-of-view. In Chapter 6 we achieve significant gains in low-delay transmission systems by use of more than $M = 8$ signal points, especially for E_s/N_0 in the range of 5...10 dB. The reason is that information theory, although it provides useful upper bounds, usually has little to say about the best *achievable* performance of low-delay transmission systems.

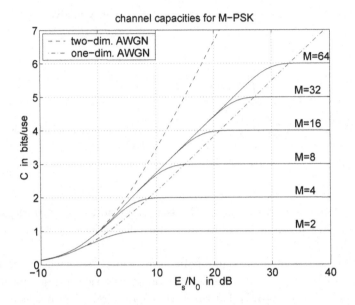

Fig. A.14   Capacities of AWGN channels with M-PSK inputs and with continuous one- and two-dimensional Gaussian inputs.

A.4   Performance Limits for the Transmission of Sources with Continuous-Values over Noisy Channels

In this section we present theoretical performance limits for the transmission of continuous-valued source signals over noisy channels. In principle, we have to appropriately combine the theoretical results from rate-distortion theory and channel capacity. At a first glance, we simply have to evaluate the DRF at a rate that equals the channel capacity, but there is an issue: the channel capacity is given in bits per channel-use but the rate we use in the DRF is given in bits per source sample. Hence, we have to scale the capacity by the number K of channel-uses with which we transmit a number N of source samples. Hence, we obtain the lowest distortion possible—the optimal performance theoretically attainable (OPTA)—by

$$D(R)|_{R=KC/N} = D\left(\frac{KC}{N}\right). \qquad (A.79)$$

In what follows we give some examples.

A.4.1 *Gaussian Source/AWGN Channel and BSC*

As a first example, we determine the performance limits for the transmission of an uncorrelated Gaussian source over an AWGN channel with continuous input alphabet and over a binary symmetric channel. The latter results from hard decisions at the output of an AWGN channel with binary input alphabet. We assume that the channel is used K times to transmit N source samples.

The DRF (A.2) for an uncorrelated Gaussian source, evaluated at the rate $R = \frac{K}{N} \cdot C$, is given by

$$D = 2^{-2\frac{K}{N} \cdot C} \cdot \sigma_x^2 \,. \tag{A.80}$$

AWGN Channel: If we insert the channel capacity (A.64) of the AWGN channel into (A.80) we obtain

$$\frac{\sigma_x^2}{D} = 2^{2K/N \cdot \frac{1}{2}\log_2(1+2E_s/N_0)} = \left(1 + \frac{2E_s}{N_0}\right)^{K/N} \,, \tag{A.81}$$

which, in terms of SNR, leads to

$$\frac{SNR}{\mathrm{dB}} = \frac{K}{N} \cdot 10\log_{10}\left(1 + \frac{2E_s}{N_0}\right) \,. \tag{A.82}$$

BSC: If we insert the channel capacity (A.39) into (A.80) we obtain

$$\frac{\sigma_x^2}{D} = 2^{2K/N \cdot (1-H(p))} \tag{A.83}$$

which, in terms of SNR, equals

$$\frac{SNR}{\mathrm{dB}} = \frac{K}{N} \cdot \underbrace{20\log_{10}(2)}_{6.02} \cdot \underbrace{(1 - H(p))}_{C} \,, \tag{A.84}$$

with the binary entropy function $H(p) = -p\log_2(p) - (1-p)\log_2(1-p)$. The bit error probability at the hard-decided AWGN channel output is $p = \frac{1}{2}\,\mathrm{erfc}(\sqrt{E_s/N_0})$ [Proakis (1995); Haykin (2000)].

The performance curves resulting from (A.82) and (A.84) are depicted in Figure A.15 for several ratios of K and N. It is obvious from (A.82) and (A.84) that the SNR-value (in dB) is doubled, when the number of channel uses per source sample is doubled. As a fundamental difference between the two channels, the SNR-values saturate only for the BSC, due to the constraint channel input alphabet that forces us to represent the source by a limited bit rate; the resulting SNR for high E_s/N_0 follows directly from the DRF for a given rate. In contrast to that, the SNR-value for the AWGN

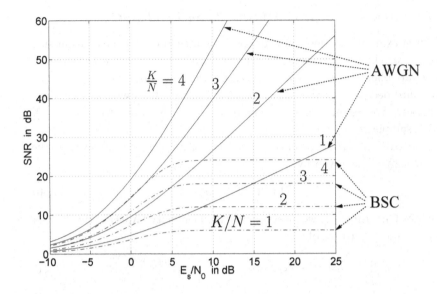

Fig. A.15 Performance limits for the transmission of an uncorrelated Gaussian source over a discrete-time AWGN channel with power-constraint and a binary symmetric channel (BSC) with bit error probability $p = \frac{1}{2}\,\mathrm{erfc}(\sqrt{E_s/N_0})$. The fraction K/N represents the number of channel uses that are used to transmit a source sample.

channel becomes infinitely large as E_s/N_0 grows to infinity, as we have no rate limitation due to the continuous channel input alphabet.

It should be mentioned that for the AWGN channel and the special case $K = N$ a very simple realizable system exists that achieves the theoretical performance limits. This system is discussed in Section 2.5.2.

A.4.2 *Correlated Gaussian Source/AWGN Channel with Binary Input*

In this section we investigate the performance limits for auto-correlated Gaussian sources that are transmitted over a binary-input AWGN channel and a BSC.

While the channel capacity for the BSC is analytically given by (A.39), we have to numerically evaluate (A.75) to obtain the channel capacity of the binary input AWGN channel for a given E_s/N_0. The capacity curves have already been determined above and they are depicted in Figure A.13.

In general, the DRF of a correlated Gaussian source is only given parametrically by (A.13) and (A.14); a specific rate-distortion pair has to be determined numerically. In our example we use the linearly filtered Gaussian source from Figure A.3, with the filter $H(z) = \frac{z}{z-a}$, $a = 0.9$,. For this case we already know the DRF from Figure A.4.

In Figure A.16 we show the best possible performance (in terms of source SNR) obtained by application of (A.79) versus the channel quality E_s/N_0 for both channels and for $K/N = 10$ and $K/N = 6$ channel-uses per source sample. For comparison we have also included the "OPTA"-curve[3] for $\frac{K}{N} = 6$ and an uncorrelated source. Due to the discrete input alphabets

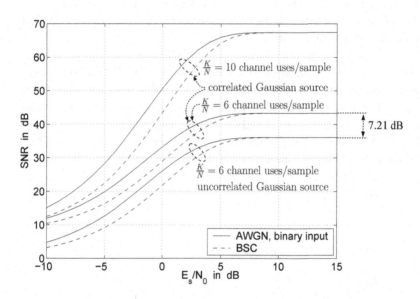

Fig. A.16 Performance limits for the transmission of a correlated first-order Gauss-Markov source with $a = 0.9$ over (i) a discrete-time AWGN channel with power-constraint and binary input and (ii) a binary symmetric channel (BSC). The ratio K/N stands for the number of channel-uses per source sample.

of both channels, the SNR-plots saturate for high E_s/N_0. The ultimate performance for very high E_s/N_0 can be easily derived from (A.38) and Figure A.13: as the channel capacity is "one" bit per use for high E_s/N_0, the bit rate is given by $R = K/N$ bits per sample, which, for $K/N = 10$, gives an SNR of $6.02 \cdot 10 + 10 \log_{10}(\frac{1}{1-a^2}) = 67.41$ dB. The additive term of 7.21 dB is due to the auto-correlation of the source. As the high-rate

---

[3]optimal performance theoretically attainable

approximation for the DRF (see Section A.2.3) works well down to a rate R of 1 bit per source sample, we can use it for $K/N = 10$ and $K/N = 6$ down to $E_s/N_0 = -5$ dB, as the capacity of both channels (see Figure A.13) is still higher than 0.2 bits per channel use, so $R = \frac{K}{N} \cdot C > 1$ bit per sample.

A.4.3 Gaussian Source/Two-Dimensional AWGN Channel with PSK Input

In this section we give the theoretical performance limits for the transmission of an uncorrelated Gaussian source over a two-dimensional AWGN channel with PSK input alphabet.

For simplicity we restrict ourselves to $K/N = 1$, i.e., we spend one channel-use for the transmission of a source sample. The DRF of the source (SNR version) is given by (A.3), and we again have to insert $R = \frac{K}{N} \cdot C$ for the rate. The channel capacity (A.78) has to be computed by numerical integration; some results are already available from Figure A.14.

Figure A.17 shows the OPTA-curves for several sizes M of the PSK modulation signal set. As above we observe a saturation of the SNR-value

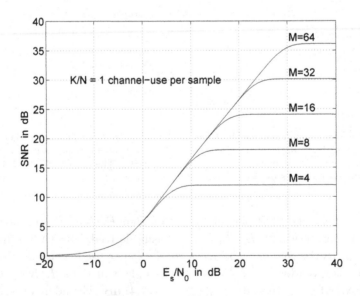

Fig. A.17 Performance limits for the transmission of Gaussian source over an AWGN channel with M-PSK input alphabet.

for high E_s/N_0 due to the constraints on the input alphabet. Furthermore, it is interesting to note that from an information-theoretic point-of-view the use of more than $M = 8$ signal points does not lead to an increase in performance if E_s/N_0 is lower than 8 dB. In Chapter 6 we show, however, that this is not true for low-delay systems. Qualitatively, very similar results as in Figure A.17 hold for M-ASK modulation signal sets and a one-dimensional Gaussian channel.

Appendix B

Optimal Decoder for a Given Encoder

At the decoder we know the received signal $\tilde{z}$ exactly, but the input vector x of the encoder just in statistical terms. Our goal here is to find an optimal decoder mapping $\varrho^{\circledast}(\tilde{z}, \varepsilon)$ for a fixed encoder ε, i.e., we want to minimize the distortion $D_D(\tilde{z}, \varepsilon, \varrho)$ defined in (2.10) for each possible "value" of $\tilde{z}$ by a proper selection of the decoder output:

$$\tilde{x}^{\circledast} = \varrho^{\circledast}(\tilde{z}, \varepsilon) = \arg\min_{\varrho} D_D(\tilde{z}, \varepsilon, \varrho) \ . \tag{B.1}$$

Since the distance measure $d(x, \tilde{x})$ in (2.7) is (by definition) always non-negative (for any combination of $x, \tilde{x}$) the total system distortion $D(\varepsilon, \varrho)$ is also minimized, if $D_D(\tilde{z}, \varepsilon, \varrho)$ is minimized for each particular "value" of the channel output $\tilde{z}$.

With the mean-squared error as a distance measure $d(\cdot, \cdot)$ in the source signal space we obtain from (2.10):

$$D_D(\tilde{z}, \varepsilon, \varrho) = \int_{\mathcal{X}} \frac{1}{N} \sum_{i=0}^{N-1} (x_i - \tilde{x}_i)^2 \cdot p(x_0, ..., x_{N-1} \mid \tilde{z}) dx_0 ... dx_{N-1} \tag{B.2}$$

with $x \doteq \{x_0, x_1, ..., x_{N-1}\}$. For the optimal choice of the components of $\tilde{x}$, we compute the partial derivatives

$$\frac{\partial D_D(\tilde{z}, \varepsilon, \varrho)}{\partial \tilde{x}_j} = \int_{\mathcal{X}} \frac{-2}{N} (x_j - \tilde{x}_j) \cdot p(x_0, ..., x_{N-1} \mid \tilde{z}) dx_0 ... dx_{N-1} \ . \tag{B.3}$$

If (B.3) is set to zero we obtain the optimal vector components

$$\tilde{x}_j^{\circledast} = \int_{\mathcal{X}} x_j \cdot p(x_0, ..., x_{N-1} \mid \tilde{z}) dx_0 ... dx_{N-1} \ , \quad j = 0, 1, ..., N-1 \ , \tag{B.4}$$

of the optimal decoder, which may be also written more compact in vector notation according to

$$\tilde{x}^{\circledast} = \int_{\mathcal{X}} x \, p(x \mid \tilde{z}) dx \ . \tag{B.5}$$

This is the well-known result for the minimum mean-square estimator, which is sometimes written as the conditional expectation

$$\tilde{x}^{\circledast} = \mathrm{E}_{X|\tilde{z}}\{X \mid \tilde{z}\} \ . \tag{B.6}$$

We can use the Bayes rule to compute the PDF

$$p(x \mid \tilde{z}) = \frac{p(\tilde{z} \mid z = \varepsilon(x)) \cdot p(x)}{p(\tilde{z})} \ , \tag{B.7}$$

where the deterministic encoder mapping ε is used, and by insertion into (B.5) we obtain:

$$\tilde{x}^{\circledast} = \frac{1}{p(\tilde{z})} \int_{\mathcal{X}} x \cdot p(\tilde{z} \mid z = \varepsilon(x)) \cdot p(x) dx \ . \tag{B.8}$$

The term $p(\tilde{z})$ is just a scaling factor that does not depend on x, because the received value $\tilde{z}$ is assumed to be known. Since the denominator can be computed by

$$\frac{1}{A} \doteq p(\tilde{z}) = \int_{\mathcal{X}} p(\tilde{z}, x) dx = \int_{\mathcal{X}} p(\tilde{z} \mid z = \varepsilon(x)) \cdot p(x) \, dx \ , \tag{B.9}$$

we finally obtain

$$\tilde{x}^{\circledast} = \varrho^{\circledast}(\tilde{z}, \varepsilon) = A \cdot \int_{\mathcal{X}} x \cdot p(\tilde{z} \mid z = \varepsilon(x)) \cdot p(x) \, dx \ . \tag{B.10}$$

To compute the receiver output vector $\tilde{x}^{\circledast}$ by (B.10) we need to know:

- the PDF $p(x)$ of the input source vectors
- the conditional PDF $p(\tilde{z} \mid z)$ of the channel output $\tilde{z}$, given the channel input z
- the received channel output vector $\tilde{z}$ corresponding to the channel-input vector z that was used to transmit the source vector x to be estimated at the receiver
- the encoder mapping $\varepsilon(x)$.

For *any* choice of the encoder mapping $\varepsilon(x)$, (B.10) defines the optimal decoder which minimizes the expected mean-squared error (B.2). The result is true, even if the source signal or the channel are correlated, as long as we use (B.2) as an optimization criterion, since the latter implicates the assumption of memoryless source and channel.

Appendix C

Symbol Error Probabilities for M-PSK

The following formula (e.g., [Haykin (2000); Proakis (1995); Benedetto and Biglieri (1999)]) for the symbol error probability p_s of M-PSK modulation and a symbol transmission over an AWGN channel is frequently used:

$$p_s \approx \mathrm{erfc}\left(\sqrt{E_s/N_0} \cdot \sin\left(\pi/M\right) \right) . \tag{C.1}$$

In (C.1), we do *not* care for the type of symbol error (direct neighbor or not) that occurs; we just say that if a channel output is located outside the decision region of the transmitted signal point (i.e., it is located in one of the half-planes in Figure 6.4), then we have a symbol error. The approximation in (C.1) is that the decision regions are changed to half-planes (instead of "cake-pieces") in order to simplify the integrations over the noise probability density functions. Hence, the "cross-hatched" region in Figure 6.4 is "counted" twice. In a comparison of simulations and results from (C.1) in Figure C.1, we observe that the approximation works very well, even for high channel noise variances. This is due to the e^{-x^2} functions involved in the probability density functions that have very small magnitudes in the "cross-hatched" region in Figure 6.4.

In the derivation of source-adaptive modulation (SAM) we assume that only direct-neighbor symbol errors occur. To check the justification of this assumption, we also plotted the direct-neighbor symbol error probabilities in Figure (C.1). We observe that at low channel SNR (depending on the size M of the signal set), the direct-neighbor errors are not the majority of all symbol errors as it is the case at high channel SNR. In Figure C.2 we show the relative portion of the direct-neighbor errors: we observe that more than 70% of the errors are direct-neighbor errors for $10\log_{10}(E_s/N_0) = 0, 5, 10$ dB for 8-, 16-, and 32-PSK, respectively. Hence, for high values of E_s/N_0 our assumption is justified and for lower values of E_s/N_0 it is good enough for our needs, which is also shown by performance simulations.

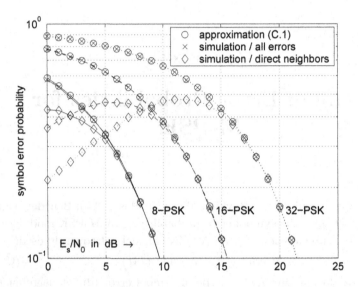

Fig. C.1   Comparison of the symbol error probabilities resulting from the approximative formula (C.1) and from simulations for 8-PSK, 16-PSK, and 32-PSK modulation.

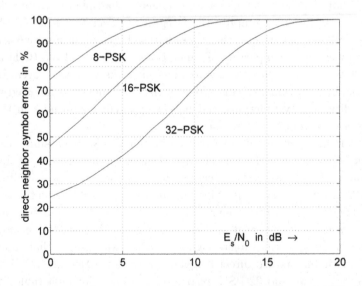

Fig. C.2   Relative number of direct-neighbor symbol errors for several PSK schemes.

Appendix D

Derivative of the Expected Distortion for SAM

We are interested to compute the derivative of (6.37) for $\Delta\varphi$, and we want to show that (6.40) holds.

First, we introduce the abbreviation

$$a_j \doteq d(x, y_{\gamma^{-1}(j)}) \quad , \tag{D.1}$$

which we insert into (6.37), resulting in

$$D(x) = a_1 + (a_0 - a_1) \cdot p'_s(\Delta\varphi) + (a_2 - a_1) \cdot p'_s(-\Delta\varphi) . \tag{D.2}$$

Since, generally,

$$\frac{d}{dx} \int_a^{u(x)} f(t)dt = f\big(u(x)\big) \cdot \frac{du(x)}{dx} \quad \text{with} \quad a = \text{const.} , \tag{D.3}$$

we obtain from (6.38) with $E_s/N_0 = \frac{1}{2\sigma_n^2}$:

$$\frac{dp'_s(\Delta\varphi)}{d\Delta\varphi} = \frac{d}{d\Delta\varphi}\left(\frac{1}{2} \, \text{erfc}\left(\frac{\sin\left(\frac{\pi}{M}+\Delta\varphi\right)}{\sqrt{2}\sigma_n} \right) \right) \tag{D.4}$$

$$= \frac{1}{2}\frac{2}{\sqrt{\pi}} \frac{d}{d\Delta\varphi}\left(\int\limits_{\frac{\sin(\frac{\pi}{M}+\Delta\varphi)}{\sqrt{2}\sigma_n}}^{+\infty} e^{-t^2} dt \right) \tag{D.5}$$

$$= -\frac{\cos\left(\frac{\pi}{M}+\Delta\varphi\right)}{\sqrt{2}\sqrt{\pi}\sigma_n} \cdot e^{-\frac{1}{2\sigma_n^2}\sin^2\left(\frac{\pi}{M}+\Delta\varphi\right)} . \tag{D.6}$$

With the substitution $v = -\Delta\varphi$ we, similarly, obtain

$$\frac{dp'_s(-\Delta\varphi)}{d\Delta\varphi} = \frac{dp'_s(v)}{dv} \cdot \frac{dv}{d\Delta\varphi} \tag{D.7}$$

$$= \frac{\cos\left(\frac{\pi}{M}-\Delta\varphi\right)}{\sqrt{2}\sqrt{\pi}\sigma_n} \cdot e^{-\frac{1}{2\sigma_n^2}\sin^2\left(\frac{\pi}{M}-\Delta\varphi\right)} . \tag{D.8}$$

Hence,

$$\frac{dD(x)}{d\Delta\varphi} = -(a_0 - a_1)\frac{\cos(\frac{\pi}{M}+\Delta\varphi)}{\sqrt{2}\sqrt{\pi}\sigma_n} \cdot e^{-\frac{1}{2\sigma_n^2}\sin^2(\frac{\pi}{M}+\Delta\varphi)} \tag{D.9}$$

$$+(a_2 - a_1)\frac{\cos(\frac{\pi}{M}-\Delta\varphi)}{\sqrt{2}\sqrt{\pi}\sigma_n} \cdot e^{-\frac{1}{2\sigma_n^2}\sin^2(\frac{\pi}{M}-\Delta\varphi)} \ . \tag{D.10}$$

If we set $\frac{dD}{d\Delta\varphi}$ to zero to find the location of the minimum of $D(x)$ we find

$$(a_0 - a_1)\frac{\cos(\frac{\pi}{M}+\Delta\varphi^\circledast)}{\sqrt{2}\sqrt{\pi}\sigma_n} \cdot e^{-\frac{1}{2\sigma_n^2}\sin^2(\frac{\pi}{M}+\Delta\varphi^\circledast)} \tag{D.11}$$

$$= (a_2 - a_1)\frac{\cos(\frac{\pi}{M}-\Delta\varphi^\circledast)}{\sqrt{2}\sqrt{\pi}\sigma_n} \cdot e^{-\frac{1}{2\sigma_n^2}\sin^2(\frac{\pi}{M}-\Delta\varphi^\circledast)} \ , \tag{D.12}$$

which, due to $a_2 - a_1 > 0$ and $\cos(\frac{\pi}{M}+\Delta\varphi) \neq 0, -\pi/M < \Delta\varphi < \pi/M$, $M \geq 4$, can be reformulated as

$$\frac{a_0 - a_1}{a_2 - a_1} = \frac{\cos(\frac{\pi}{M}-\Delta\varphi^\circledast)}{\cos(\frac{\pi}{M}+\Delta\varphi^\circledast)} \cdot e^{\frac{1}{2\sigma_n^2}\left(\sin^2(\frac{\pi}{M}+\Delta\varphi^\circledast)-\sin^2(\frac{\pi}{M}-\Delta\varphi^\circledast)\right)} \ . \tag{D.13}$$

The exponent on the right-hand side in (D.13) can be rewritten as follows:

$$\sin^2(\frac{\pi}{M}+\Delta\varphi^\circledast) - \sin^2(\frac{\pi}{M}-\Delta\varphi^\circledast)$$

$$= \frac{1}{2}\left(1 - \cos(2\frac{\pi}{M}+2\Delta\varphi^\circledast)\right) - \frac{1}{2}\left(1 - \cos(2\frac{\pi}{M}-2\Delta\varphi^\circledast)\right)$$

$$= -\frac{1}{2}\left(\cos(2\frac{\pi}{M}+2\Delta\varphi^\circledast) - \cos(2\frac{\pi}{M}-2\Delta\varphi^\circledast)\right) \tag{D.14}$$

By insertion of

$$\cos(\frac{2\pi}{M} \pm 2\Delta\varphi^\circledast) = \cos\left(\frac{2\pi}{M}\right)\cos(2\Delta\varphi^\circledast) \mp \sin\left(\frac{2\pi}{M}\right)\sin(2\Delta\varphi^\circledast) \ , \tag{D.15}$$

(D.14) can be simplified to

$$\sin^2\left(\frac{\pi}{M} + \Delta\varphi^\circledast\right) - \sin^2\left(\frac{\pi}{M} - \Delta\varphi^\circledast\right) = \sin\left(\frac{2\pi}{M}\right) \cdot \sin(2\Delta\varphi^\circledast) \ . \tag{D.16}$$

If we insert (D.16) into (D.13) we obtain (6.40).

List of Symbols

a	filter coefficient of a first-order Markov model
b_k	bit-vector at time k (e.g. a quantizer output)
$b_{k,n}$	n-th bit of b_k, $b_{k,n} \in \{0,1\}$, $n = 1, 2, ..., N_B$
$\mathcal{B}$	set of all possible N_B-bit vectors, i.e., $\mathcal{B} = \{0,1\}^{N_B}$
c_j	centroid of quantizer partition region number j in COVQ
C	channel capacity in bits per channel-use
$d(x, \tilde{x})$	measure for the distance between the vectors x and $\tilde{x}$
$d_H(i, j)$	Hamming distance between the bit-vector representations of the indexes i, j. This is a short version of $d_H(\gamma(i), \gamma(j))$, with γ the natural binary bit mapping)
D	expected value of a distortion
$\mathrm{erfc}(x)$	$= \frac{2}{\sqrt{\pi}} \int_x^{+\infty} \exp(-t^2)dt$ (complementary error function)
E_l	bit energy
E_s	modulation symbol energy
E	expectation
f	frequency in Hz
f_S	sampling frequency
g_i	quantizer decision boundary
$h(X)$	differential entropy of a continuous-valued random variable X
$H(I)$	entropy of a quantizer index I
$H(I, J)$	joint entropy of the quantizer indexes I, J
i, j	quantizer output indexes, realisations of random variables I, J
$\mathcal{I}, \mathcal{J}$	sets of all possible indexes i, j
k	discrete time index

K	number of channel uses or number of multiple descriptions to transmit an N-dimensional source vector
l	index for vector components
M	number of modulation signal points
N	dimension of a source vector
N_B	number of bits to represent a quantizer index
$N_Y \doteq 2^{N_B}$	number of code-vectors in a codebook of an N_B bit quantizer
$N_0/2$	noise power spectral density on a continuous-time Gaussian channel
$\mathcal{N}(0,1)$	normal (Gaussian) distribution with zero mean and unit variance
p_e	bit error probability
p_s	symbol error probability
P	transmission power
$\mathrm{P}(j)$	probability of the index j
$\mathrm{P}_{j|i}$	probability that the quantizer index j is received, given that i is transmitted
$q(x)$	energy of the signal vector x, i.e., $q(x) = \sum_l x_l^2$
$Q(x)$	index of the nearest neighbor of the source vector x in a quantizer codebook
R	source-coding bit rate in bits per source sample
R_c	channel code rate
$\mathrm{si}(x)$	si-function, $\mathrm{si}(x) \doteq \sin(x)/x$
s_j	signal point of a digital modulation scheme
$\mathcal{S}_M$	modulation signal constellation (or set); $\mathcal{S}_M \doteq \{s_0, s_1, ..., s_{M-1}\}$
t	continuous time variable
T_s	sampling period
w	noise signal, random variable W
$w_H(z)$	Hamming-weight of the bit-vector z (number of "1"-bits in z)
$x(t)$	continuous-time signal
x_k	N-dimensional discrete-time source vector
$\tilde{x}_k$	reproduction of the discrete-time source vector x_k at the receiver output
$X(f)$	Fourier-transform of $x(t)$

$X(\Omega)$	Fourier-transform of the discrete-time signal x_k, $\Omega \in (-\pi, \pi)$
$\mathcal{X}$	signal space of the N-dimensional input source vector: $x \in \mathcal{X} = I\!R^N$
$\mathcal{X}_j$	decision region of a quantizer for reproduction level y_j
y_i	quantizer code-vector (VQ/COVQ) at line i of the codebook
$y(b_k)$	quantizer code-vector that is coded by the bit-vector b_k; short version of $y_{\gamma^{-1}(b_k)}$
$\mathcal{Y}$	quantizer codebook
z	channel input; random variable Z
z_I	inphase component of the channel input
z_Q	quadrature component of the channel input
$\tilde{z}$	channel output; random variable $\tilde{Z}$
$\tilde{z}_I$	inphase component of the channel output
$\tilde{z}_Q$	quadrature component of the channel output
$\mathcal{Z}$	set of possible channel inputs
$\tilde{\mathcal{Z}}$	set of possible channel outputs
$\tilde{\mathcal{Z}}_j$	decision region of a hard-decision detector for the signal point s_j
$\varepsilon(x) = z$	encoder mapping
$\varrho(\tilde{z}) = \tilde{x}$	decoder mapping
$\gamma(i)$	index mapping from a quantizer index i to a modulation signal point
$\gamma(i) = b$	mapping from a quantizer output index i to a bit vector b
$\gamma^{-1}(b) = j$	mapping from a quantizer output bit-vector b to an index j
$\delta(t)$	continuous-time delta function
$\varphi_{xx}(\lambda)$	autocorrelation sequence of a discrete-time random variable x_k
$\Phi_{xx}(\Omega)$	power spectral density of the discrete-time random variable x_k
$\pi(b)$	bit interleaver
$\sigma_w^2 = \frac{N_0}{2T_s}$	channel noise variance at the sampled matched-filter output
$\sigma_n^2 = \frac{N_0}{2E_s}$	channel noise variance in the normalized channel model; the mean power of the signal constellation is assumed to be one

List of Acronyms

AbS	analysis-by-synthesis (source-coding principle)
ACF	auto-correlation function
ACS	auto-correlation sequence
APP	a-posteriori probability
AWGN	additive white Gaussian noise
BCJR	algorithm to compute APPs [Bahl *et al.* (1974)]
BFI	bad frame indication
BFH	bad frame handling
BSA	binary switching algorithm
BSC	binary symmetric channel
CCOE	channel-coded optimal estimation
CELP	code-excited linear prediction (speech coding principle)
COVQ	channel-optimized vector quantization
CRC	cyclic redundancy check
CSI	channel state information
DFT	discrete Fourier transform
DPCM	differential pulse code modulation (equivalent to AbS)
DRF	distortion-rate function
EDGE	enhanced data rates for GSM evolution
EFR	enhanced-full-rate speech codec)
ETSI	European Telecommunications Standards Institute
FFT	fast algorithm to compute the DFT
GPRS	general packet radio service
GSM	Global System for Mobile Communications
GS-VQ	gain-shape vector quantization

HD	hard decision
IDFT	inverse discrete Fourier-transform
ISCD	iterative source-channel decoding
ITU	International Telecommunication Union
JSCC	joint source-channel coding
JSCD	joint source-channel decoding
LBG	algorithm for vector quantizer codebook training [Linde *et al.* (1980)])
LPC	linear predictive coding
MD	multiple descriptions
MD-BSA	binary switching algorithm for multiple descriptions
MDVQ	multiple description vector quantization
MSE	mean-squared error
OE	optimal estimation (optimal source decoding without channel coding)
OPTA	optimal performance theoretically attainable
PDF	probability density function
PGC	pseudo-Gray-coding
PMF	probability mass function
PSD	power spectral density
RDF	rate-distortion function
SAM	source-adaptive modulation
SAP	source-adaptive power allocation
SAS	source-adaptive shifting of modulation signal sets (special version of source-adaptive modulation (SAM))
SD	soft decision
SFM	spectral flatness measure
SNR	signal-to-noise ratio in dB
SOVA	soft-output Viterbi algorithm
UEP	unequal error protection
UMTS	Universal Mobile Telecommunication System
VLC	variable-length code (e.g., Huffman code)
VQ	vector quantization

Bibliography

Alajaji, F. and Phamdo, N. (1998). Soft-decision COVQ for Rayleigh-fading channels, *IEEE Communications Letters* **2**, 6, pp. 162–164.

Alajaji, F., Phamdo, N. and Fuja, T. (1996). Channel codes that exploit the residual redundancy in CELP-encoded speech, *IEEE Transactions on Speech and Audio Processing* **4**, pp. 325–336.

Bahl, L. R., Cocke, J., Jelinek, F. and Raviv, J. (1974). Optimal decoding of linear codes for minimizing symbol error rate, *IEEE Transactions on Information Theory* **IT-20**, pp. 284–287.

Bauer, R. (2002). *Kanalcodierung für Quellencodes variabler Länge*, Ph.D. thesis, Munich University of Technology.

Ben-David, G. and Malah, D. (1994). Simple adaptation of vector-quantizers to combat channel errors, in *Proceedings of the 6-th DSP Workshop* (Yosemite, California, USA), pp. 41–44.

Benedetto, S. and Biglieri, E. (1999). *Principles of Digital Transmission* (Kluwer Academic / Plenum Publishers, New York).

Berger, T. (2003). Living information theory, *IEEE Information Theory Society Newsletter* **53**, 1, pp. 1, 6–19.

Berger, T. and Gibson, J. D. (1998). Lossy source coding, *IEEE Transactions on Information Theory* **44**, 6, pp. 2693–2723.

Berrou, C. and Glavieux, A. (1996). Near optimum error correcting coding and decoding: Turbo-codes, *IEEE Transactions on Communications* **44**, 10, pp. 1261–1271.

Cheng, N.-T. and Kingsbury, N. (1989). Robust zero-redundancy vector quantization for noisy channels, in *Proceedings of the IEEE International Conference on Communications (ICC)*, pp. 1338–1342.

Cover, T. and Thomas, J. (1991). *Elements of Information Theory* (John Wiley & Sons, Inc.).

Farvardin, N. (1990). A study of vector quantization for noisy channels, *IEEE Transactions on Information Theory* **36**, 4, pp. 799–809.

Farvardin, N. and Vaishampayan, V. (1991). On the performance and complexity of channel-optimized vector quantizers, *IEEE Transactions on Information Theory* **37**, 1, pp. 155–160.

Fazel, T. and Fuja, T. (2003). Robust transmission of MELP-compressed speech: an illustrative example of joint source-channel decoding, *IEEE Transactions on Communications* **51**, 6, pp. 973–982.

Fingscheidt, T. and Vary, P. (1997). Robust speech decoding: A universal approach to bit error concealment, in *Proceedings of the IEEE International Conference on Acoustics, Speech, and Signal Processing (ICASSP)*, Vol. 3, pp. 1667–1670.

Fleming, M. and Effros, M. (1999). Generalized multiple description vector quantization, in *Proceedings of the IEEE Data Compression Conference*, pp. 3–12.

Foschini, G. J., Gitlin, R. D. and Weinstein, S. B. (1974). Optimization of two-dimensional signal constellations in the presence of Gaussian noise, *IEEE Transactions on Communications* **COM-22**, 1, pp. 28–38.

Gadkari, S. and Rose, K. (1999). Robust vector quantizer design by noisy channel relaxation, *IEEE Transactions on Communications* **47**, 8, pp. 1113–1116.

Gallager, R. G. (1968). *Information Theory and Reliable Communication* (John Wiley & Sons, Inc.).

Gastpar, M., Rimoldi, B. and Vetterli, M. (2003). To code, or not to code: lossy source-channel communication revisited, *IEEE Transactions on Information Theory* **49**, 5, pp. 1147–1158.

Gersho, A. and Gray, R. M. (2001). *Vector Quantization and Signal Compression*, 8th edn. (Kluwer Academic Publishers).

Goertz, N. (1998). Joint source channel decoding using bit-reliability information and source statistics, in *Proceedings of the IEEE International Symposium on Information Theory (ISIT)*, p. 9.

Goertz, N. (1999). Joint source-channel decoding by channel-coded optimal estimation (CCOE) for a CELP speech codec, in *Proceedings of the European Conference on Speech Communication and Technology (EUROSPEECH)*, Vol. 3, pp. 1447–1450.

Goertz, N. (2000). Analysis and performance of iterative source-channel decoding, in *Proceedings of the International Symposium on Turbo Codes & Related Topics*, pp. 251–254.

Goertz, N. (2001a). A generalized framework for iterative source-channel decoding, *Annals of Telecommunications, Special issue on "Turbo Codes: a widespreading Technique"* , pp. 435–446.

Goertz, N. (2001b). On the iterative approximation of optimal joint source-channel decoding, *IEEE Journal on Selected Areas in Communications* **19**, 9, pp. 1662–1670.

Goertz, N. (2002). Source-adaptive shifting of modulation signal points for improved transmission of waveform-signals over noisy channels, in *Proceedings of the IEEE International Conference on Acoustics, Speech, and Signal Processing (ICASSP)*, Vol. 3, pp. 2517–2520.

Goertz, N. and Heute, U. (2000). Joint source-channel decoding with iterative algorithms, in *Proceedings of the European Signal Processing Conference (EUSIPCO)*, Vol. 1, pp. 425–428.

Goertz, N. and Kliewer, J. (2003). Memory efficient adaptation of vector quantizers to time-varying channels, *EURASIP Signal Processing* **83**, pp. 1519–1528.

Goyal, V. K., Kelner, J. A. and Kovačević, J. (2002). Multiple description vector quantization with a coarse lattice, *IEEE Transactions on Information Theory* **48**, 3, pp. 781–788.

Hagenauer, J. (1992). Soft-in/soft-out: The benefits of using soft-decisions in all stages of digital receivers, in *3rd International Workshop on DSP Techniques applied to Space Communications*.

Hagenauer, J. (1995). Source-controlled channel decoding, *IEEE Transactions on Communications* **43**, 9, pp. 2449–2457.

Hagenauer, J. and Goertz, N. (2003). The turbo principle in joint source-channel coding, in *Proceedings of the IEEE Information Theory Workshop*, pp. 275–278.

Hagenauer, J., Offer, E. and Papke, L. (1996). Iterative decoding of binary block and convolutional codes, *IEEE Transactions on Information Theory* **42**, 2, pp. 429–445.

Han, J.-K. and Kim, H.-M. (2001). Joint optimization of VQ codebooks and QAM signal constellations for AWGN channels, *IEEE Transactions on Communications* **49**, 5, pp. 816–825.

Haykin, S. (2000). *Communication Systems*, 4th edn. (John Wiley & Sons).

Heinen, S. (2001). *Quellenoptimierter Fehlerschutz für digitale Übertragungskanäle*, Ph.D. thesis, RWTH Aachen University.

Heinen, S., Geiler, A. and Vary, P. (1998). MAP channel decoding by exploiting multilevel source a priori knowledge, in *Proceedings of the ITG-Fachtagung "Codierung für Quelle, Kanal und Übertragung"*, pp. 89–94.

Jafarkhani, H. and Farvardin, N. (1996). Channel-matched hierarchical table-lookup vector quantization for transmission of video over wireless channels, in *Proceedings of the IEEE International Conference on Image Processing (ICIP)*, pp. 755–758.

Jafarkhani, H. and Farvardin, N. (2000). Design of channel-optimized vector quantizers in the presence of channel mismatch, *IEEE Transactions on Communications* **48**, 1, pp. 118–124.

Jayant, N. S. and Noll, P. (1984). *Digital Coding of Waveforms* (Prentice-Hall, Inc., Englewood Cliffs, New Jersey).

Kumazawa, H., Kasahara, M. and Namekawa, T. (1984). A construction of vector quantizers for noisy channels, *Electronics and Engineering in Japan* **67–B**, 4, pp. 39–47.

Lin, S. and Costello, D. J. (1983). *Error Control Coding* (Prentice-Hall, Inc., Englewood Cliffs, New Jersey).

Linde, Y., Buzo, A. and Gray, R. M. (1980). An algorithm for vector quantizer design, *IEEE Transactions on Communications* **COM-28**, 1, pp. 84–95.

Liu, F.-H., Ho, P. and Cuperman, V. (1993). Joint source and channel coding using a non-linear receiver, in *Proceedings of the IEEE International Conference on Communications (ICC)*, Vol. 3, pp. 1502–1507.

Masnick, B. and Wolf, J. (1967). On linear unequal error protection codes, *IEEE Transactions on Information Theory* **IT-3**, 4, pp. 600–607.

McEliece, R. J. (2002). *The Theory of Information and Coding*, 2nd edn. (Cambridge University Press, Cambridge).

Miller, D. and Rose, K. (1994). Combined source-channel vector quantization using deterministic annealing, *IEEE Transactions on Communications* **42**, 2/3/4, pp. 347–356.

Miller, D. J. and Park, M. (1998). A sequence-based approximate MMSE decoder for source coding over noisy channels using discrete hidden Markov models, *IEEE Transactions on Communications* **46**, 2, pp. 222–231.

Ozarow, L. (1980). On a source-coding problem with two channels and three receivers, *The Bell System Technical Journal* **59**, 10, pp. 1909–1921.

Park, M. and Miller, D. J. (2000). Joint source-channel decoding for variable-length encoded data by exact and approximate MAP sequence estimation, *IEEE Transactions on Communications* **48**, 1, pp. 1–6.

Perkert, R., Kaindl, M. and Hindelang, T. (2001). Iterative source and channel decoding for GSM, in *Proceedings of the IEEE International Conference on Acoustics, Speech, and Signal Processing (ICASSP)*, Vol. 4, pp. 2649–2652.

Phamdo, N. and Farvardin, N. (1994). Optimal detection of discrete Markov sources over discrete memoryless channels—applications to combined source-channel coding, *IEEE Transactions on Information Theory* **40**, 1, pp. 186–193.

Proakis, J. G. (1995). *Digital Communications*, 3rd edn. (McGraw-Hill International Editions).

Robertson, P., Hoeher, P. and Villebrun, E. (1997). Optimal and sub-optimal maximum a posteriori algorithms suitable for turbo decoding, *European Transactions on Telecommunications (ETT)* **8**, 2, pp. 119–125.

Sayood, K. and Borkenhagen, J. C. (1991). Use of residual redundancy in the design of joint source/channel coders, *IEEE Transactions on Communications* **39**, 6, pp. 838–846.

Schimmel, M. (2003). *Anwendung der quellenadaptiven Modulation auf die AMR-Sprachcodecs*, Diploma thesis, Institute for Communications Engineering, Munich University of Technology.

Schreckenbach, F., Goertz, N., Hagenauer, J. and Bauch, G. (2003). Optimized symbol mappings for bit-interleaved coded modulation with iterative decoding, in *Proceedings of the IEEE Global Communications Conference (GLOBECOM)*.

Schroeder, M. R. and Atal, B. S. (1985). Code-excited linear prediction (CELP): high-quality speech at very low bit rates, in *Proceedings of the IEEE International Conference on Acoustics, Speech, and Signal Processing (ICASSP)*, Vol. 1, pp. 25.1.1–25.1.4.

Schuessler, H. W. (1991). *Netzwerke, Signale und Systeme, Band 2*, 3rd edn. (Springer-Verlag).

Skoglund, M. (1999). Soft decoding for vector quantization over noisy channels with memory, *IEEE Transactions on Information Theory* **45**, 4, pp. 1293–1307.

Vaishampayan, V. A. (1993). Design of multiple description scalar quantizers, *IEEE Transactions on Information Theory* **39**, 3, pp. 821–834.

Vaishampayan, V. A. and Farvardin, N. (1992). Joint design of block source codes and modulation signal sets, *IEEE Transactions on Information Theory* **38**, 4, pp. 1230–1248.

Vaishampayan, V. A., Sloane, N. J. A. and Servetto, S. D. (2001). Multiple-description vector quantization with lattice codebooks: Design and analysis, *IEEE Transactions on Information Theory* **47**, 5, pp. 1718–1734.

Zeger, K. and Gersho, A. (1988). Vector quantizer design for memoryless noisy channels, in *Proceedings of the IEEE International Conference on Communications (ICC)*, Vol. 3, pp. 1593–1597.

Zeger, K. and Gersho, A. (1990). Pseudo-Gray coding, *IEEE Transactions on Communications* **38**, 12, pp. 2147–2158.

Zimmermann, H.-M. (2003). *Iterative Source-Channel Decoding with Turbo Codes*, Diploma thesis, Institute for Communications Engineering, Munich University of Technology.

Index